anatomy of
YOGA

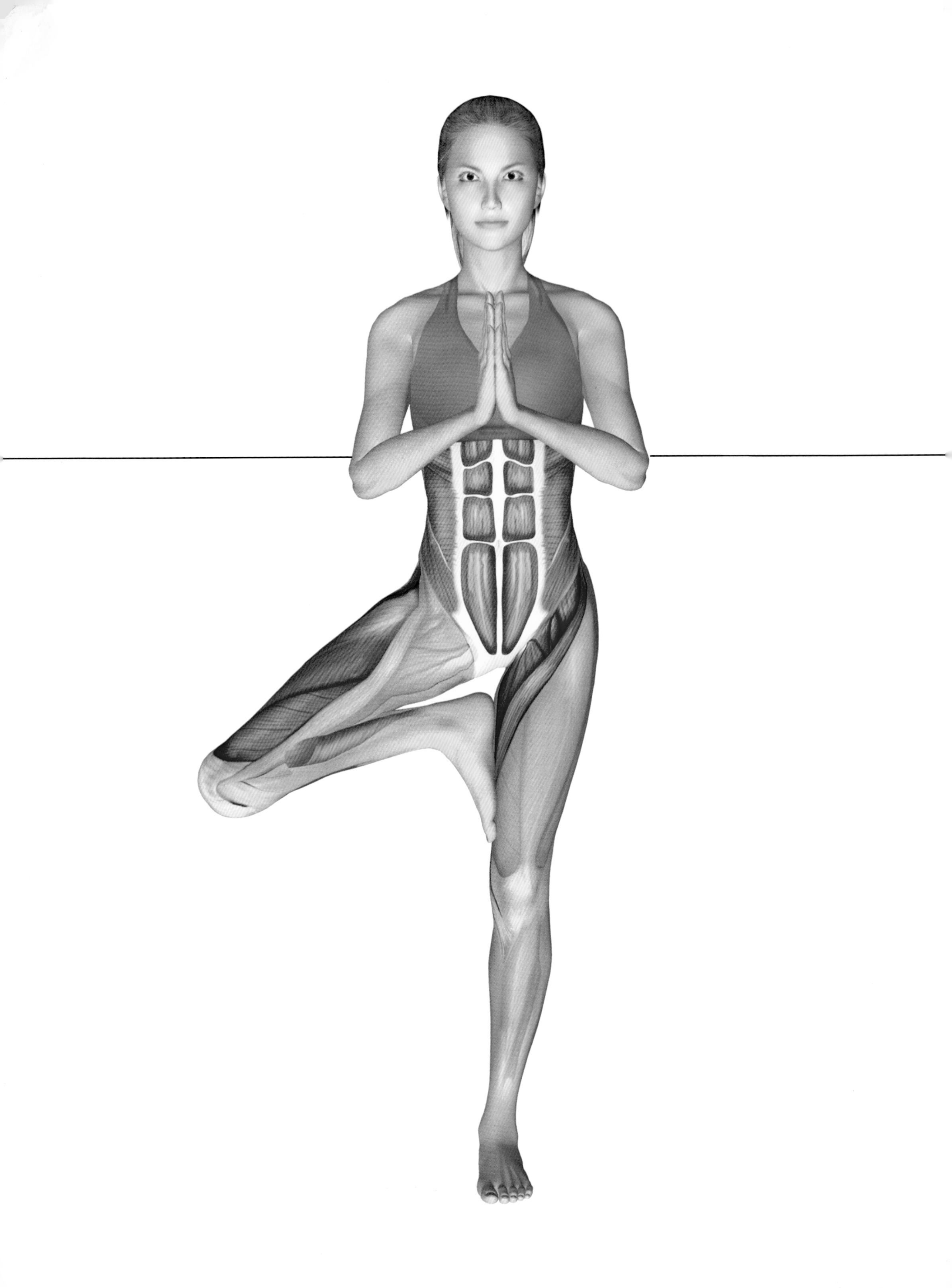

anatomy of YOGA

Dr. Abigail Ellsworth

Firefly Books

General Disclaimer

The contents of this book are intended to provide useful information to the general public. All materials, including texts, graphics, and images, are for informational purposes only and are not a substitute for medical diagnosis, advice, or treatment for specific medical conditions. All readers should seek expert medical care and consult their own physicians before commencing any exercise program or for any general or specific health issues. The author and publishers do not recommend or endorse specific treatments, procedures, advice, or other information found in this book and specifically disclaim all responsibility for any and all liability, loss, or risk, personal or otherwise, which is incurred as a consequence, directly or indirectly, of the use or application of any of the material in this publication.

A FIREFLY BOOK

Published by Firefly Books Ltd. 2010

First printing

Publisher Cataloging-in-Publication Data (U.S.)

Ellsworth, Abigail.
Anatomy of yoga : an instructor's inside guide to improving your poses / Abigail Ellsworth.
[160] p. : col. ill., col. photos. ; cm.
Summary: Step-by-step instructions and key muscles used for yoga poses.
ISBN-13: 978-1-55407-698-7 (bound) ISBN-13: 978-1-55407-766-3 (pbk.)
ISBN-10: 1-55407-698-6 (bound) ISBN-10: 1-55407-766-4 (pbk.)
1. Human anatomy. 2. Yoga. 3. Anatomy. 4. Physiology.I. Title.
613.7/046 dc22 RA781.7.E557 2010

Library and Archives Canada Cataloguing in Publication

Ellsworth, Abigail
Anatomy of yoga : an instructor's inside guide to improving your poses / Abigail Ellsworth.
ISBN-13: 978-1-55407-698-7 (bound) ISBN-13: 978-1-55407-766-3 (pbk.)
ISBN-10: 1-55407-698-6 (bound) ISBN-10: 1-55407-766-4 (pbk.)
1. Hatha yoga. I. Title.
RA781.7.E55 2010 613.7'046 C2010-90985-7

Published in the United States by
Firefly Books (U.S.) Inc.
P.O. Box 1338, Ellicott Station
Buffalo, New York 14205

Published in Canada by
Firefly Books Ltd.
66 Leek Crescent
Richmond Hill, Ontario L4B 1H1

Printed in China

Anatomy of Yoga was developed by:
Moseley Road Inc.
123 Main Street
Irvington, New York 10533

Moseley Road Inc.
President: Sean Moore
International Rights Director: Karen Prince
Art Director: Brian MacMullen
Editorial Director: Lisa Purcell

Designers: Amy Pierce, Terasa Bernard
Production: Hwaim Holly Lee
Assistant Editor: Jon Derengowski
Contributing Writers: Amy Pierce, Lisa Purcell

Photographer: Jonathan Conklin Photography, Inc.
Model: Zahava "Goldie" Karpel

CONTENTS

CONTENTS continued

INTRODUCTION

The practice of yoga, developed in India thousands of years ago, aims to educate the body, mind, and spirit. Today, this ancient system has become one of the most popular ways of both getting fit and finding some serenity in today's hectic world. Through breathing techniques and perfecting a series of poses—known as *asanas*—students of yoga refresh both body and spirit.

ANATOMY OF YOGA focuses on the physical aspect of yoga and features more than fifty asanas common to many yoga disciplines. Step-by-step photos and anatomical illustrations guide you through attaining the asanas, with the muscles strengthened in each pose highlighted. There are also handy tips that guide you as you learn to achieve and hold each asana and note each asana's focus to better allow you to target certain areas of your body. The asanas are grouped into five sections—Standing, Forward Bends, Backbends, Seated Poses & Twists, and Arm Supports & Inversions—along with a section that helps you pull them altogether in flowing sequences.

BREATHING BASICS

Yoga is largely embodied by the physical poses, or asanas, that one practices. The asanas focus on strength, flexibility, and bodily control. Beneath our structure of bones, tendons, and muscles, however, there is an entire respiratory system working simultaneously. Similar to the processes in digestion and cellular function, breathing draws nutrients into your body and expels waste. Breath is the link between our physical and mental selves, and breath control, or Pranayama, is an important yoga practice that you should exercise separately and incorporate into the asana practice. Expanding and strengthening your breath and mind, then, coincides with stretching and strengthening your body.

BREATH CONTROL
(PRANAYAMA)

Practicing Pranayama means to control your internal pranic energy, or breath of life. *Apana* refers to the elimination of breath—the alternate action of *prana.* While you intake the breath of life, you must also eliminate the toxins within the depths of your respiratory system.

There are numerous Pranayama exercises for you to practice the movement of prana. Below are examples of both rejuvenating and relaxing exercises, all of them aimed at replenishing fresh oxygen to your lungs and connecting your mind with your body.

PRONUCIATION & MEANING
- Pranayama (prah-nah-YAHM-ah)
- *prana* = breath's internal energy, the breath of life; *pra* = before; *an* = to breathe, to live; *ayama* = extension, control

LEVEL
- All levels

BENEFITS
- Restores health and mental clarity
- Provides relief from stress
- Improves emotional and physical control
- Increases awareness of the body's rhythms

1 SAMAVRTTI = SAME ACTION

Observe the irregularities of your breathing, and transition into a slower and more even breath. To achieve the same action, or samavrtti, inhale for four counts, and then exhale for four counts. This breathing technique calms the mind and creates a sense of balance and stability.

2 UJJAYI = THE VICTORIOUS BREATH

Ujjayi is sometimes called the "ocean breath" because of the sound air makes as it passes through the narrowed epiglottal passage. Maintaining the same even rhythm as the breath in Samavritti, constrict your epiglottis in the back of your throat to practice Ujjayi. Keep your mouth closed, and listen for the hiss in the back of your throat. Ujjayi breathing tones internal organs, increases internal body heat, improves concentration, and calms the mind and body.

3 KUMBHAKA = RETAINING THE BREATH

Kumbhaka is the practice of holding your breath. Begin by practicing Ujjayi or Samavrtti breathing. After every four successive breaths, hold your breath in Kumbhaka for four to eight counts. Then, allow your exhalation to last longer than your inhalation. Initially, your Kumbhaka will be shorter than your other breaths. Eventually, reduce the number of breaths in between Kumbhaka breaths and increase the number of counts in your inhale, exhale, and Kumbhaka. Build up to an exhalation twice as long as your inhalation, and a Kumbhaka breath three times as long. Kumbhaka practice strengthens the diaphragm, restores energy, and cleanses the respiratory system.

BEGINNING YOUR PRANAYAMA BREATHING PRACTICE

Before practicing Pranayama in seated positions, lie down in Corpse Pose (Savasana, see page 29) to focus on your breath. Breathe evenly, and focus on filling every part of your lungs with oxygen. Air should fill your lungs from the bottom up. First, your diaphragm expands to fill your abdomen. Air then fills the middle of your lungs within your rib cage, until it finally reaches the top of your lungs, indicated by the rising of your chest. Both sides of your chest should rise equally. Most people only fill the top of their lungs with air, leaving the bottom portions largely deprived of proper nourishment. When you are ready to practice Pranayama in a comfortable seated position, begin by placing one hand on your chest and the other on your abdominals. This will help you observe your breath. Close your eyes, lift up out of your spine, tuck your chin slightly toward your sternum, and listen to your breath as your rib cage and abdominals expand and contract. Focus on the pathway that your breath travels, the rhythm, and the texture of the sound.

BREATH CONTROL CONTINUED
(PRANAYAMA)

❹ ANULOMA VILOMA = ALTERNATE NOSTRIL BREATHING

Anuloma Viloma purifies the energy channels, or *nadis*, through the right and left nostrils. This stimulates the movement of prana. Begin by forming the Vishnu Mudra hand position, with your index and middle fingers of your right hand curled down. Place your thumb on the outside of your right

❶ To form the Vishnu Mudra hand position, curl your index and middle fingers down, while keeping your ring finger and pinkie close together and pointed upward.

To stimulate Ajna chakra, place your index and middle fingers on forehead. The Ajna chakra is known as the chakra of the mind. This space between your eyebrows is said to be where the nadis energy channels through your nostrils and meets with the central nadi. This is a very powerful hand position in Pranayama practice.

❷ In the Vishnu Mudra hand position, close your right nostril with your right thumb, and inhale through your left.

❸ Hold your breath, squeezing both nostrils with your ring finger and your thumb, and then release your thumb to exale through your right nostril.

nostril, and inhale through your left nostril, keeping your mouth closed. Close your left nostril with your ring finger at the top of the breath, and hold momentarily. Lift your thumb, and exhale out of your right nostril. Then, inhale with your right nostril, and so on. Begin with five cycles, gradually increasing the number of cycles with practice. Anuloma Viloma lowers the heart rate and relieves stress.

❺ KAPALABHATI = THE SHINING SKULL

Kapalabhati breathing incorporates a rhythmic pumping action in the abdominals to exhale. Begin by loosening your abdominals, and fill your diaphragm with air. Then, push the air out of your belly in a quick, explosive exhale. The inhalation automatically follows. This is one cycle. Begin with two rounds of ten cycles and gradually build to four rounds of twenty cycles. Kapalaphati strengthens the diaphragm, restores energy, and cleanses the respiratory system.

❻ SITHALI = THE COOLING BREATH

Unlike most other Pranayama exercises, inhalation occurs through the mouth in Sithali breathing. To practice Sithali, curl the sides of your tongue, and stick it slightly outside of your mouth. Inhale through the divot of your tongue. Retain your breath, close your mouth, and exhale through your nose. Continue to five or ten cycles. Sithali cools the body, providing comfort.

Sithali breathing literally cools your body. Curl your tongue, and inhale through your mouth.

UPPER BODY
(FRONT)

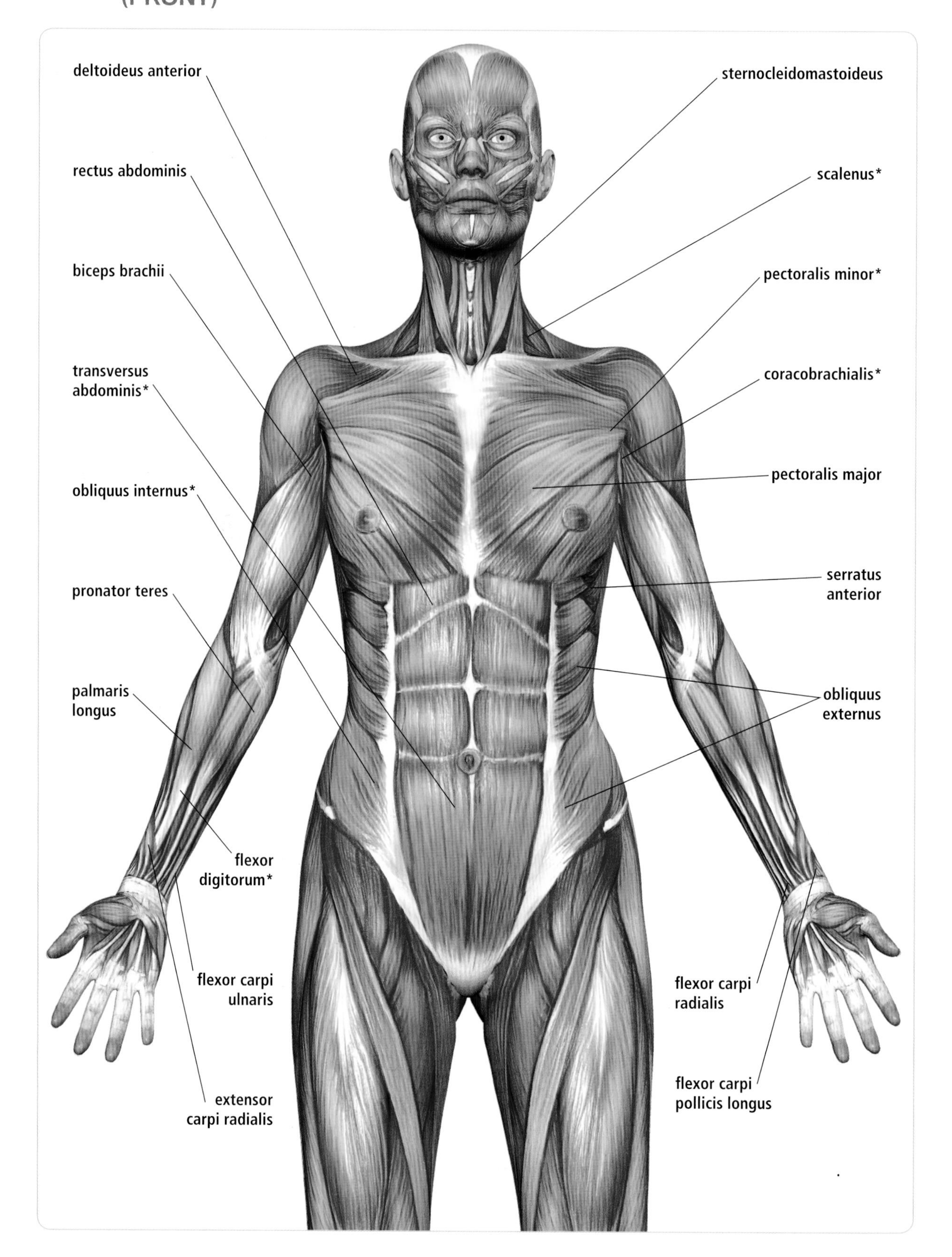

UPPER BODY
(BACK)

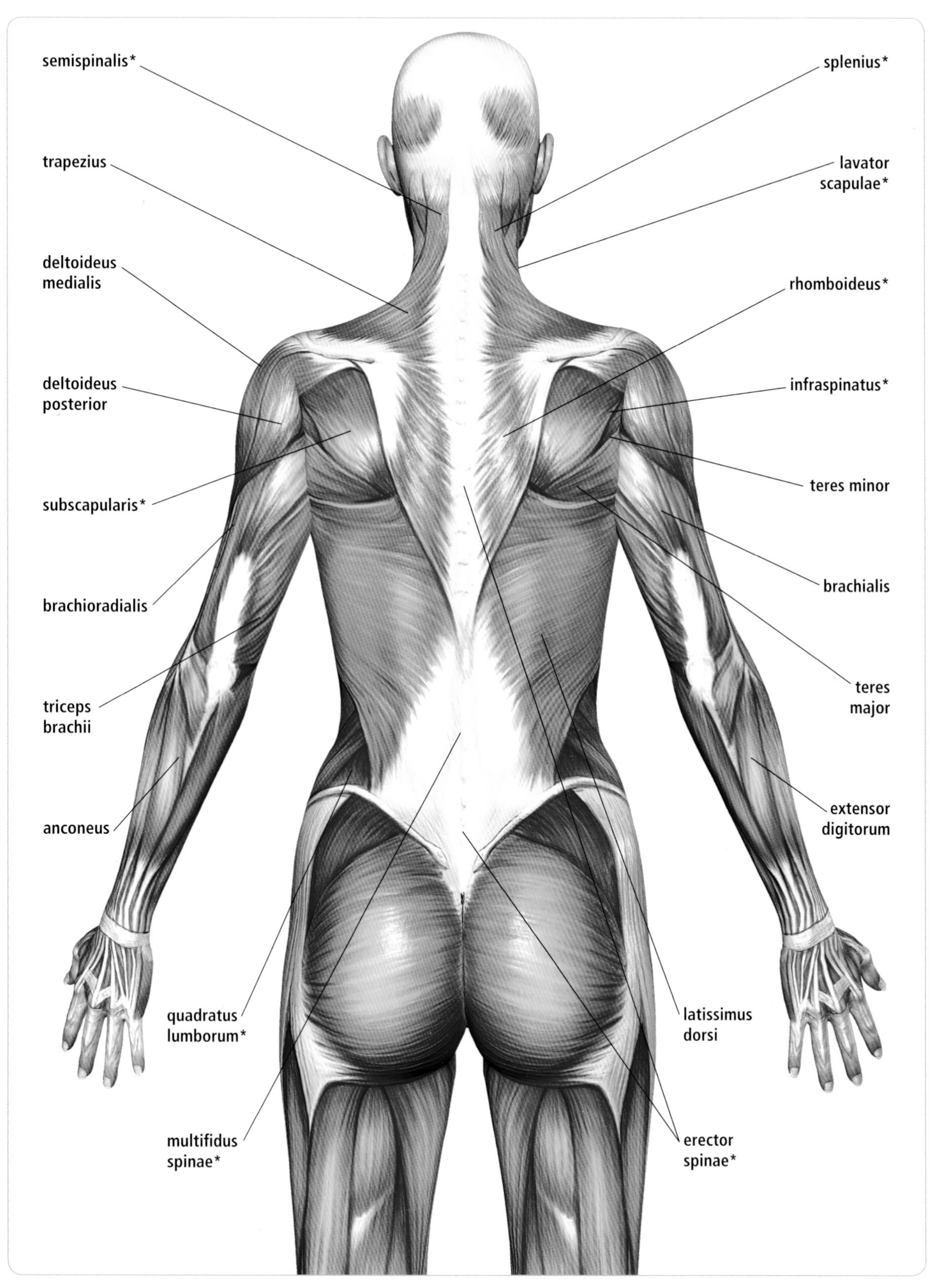

LOWER BODY
(FRONT)

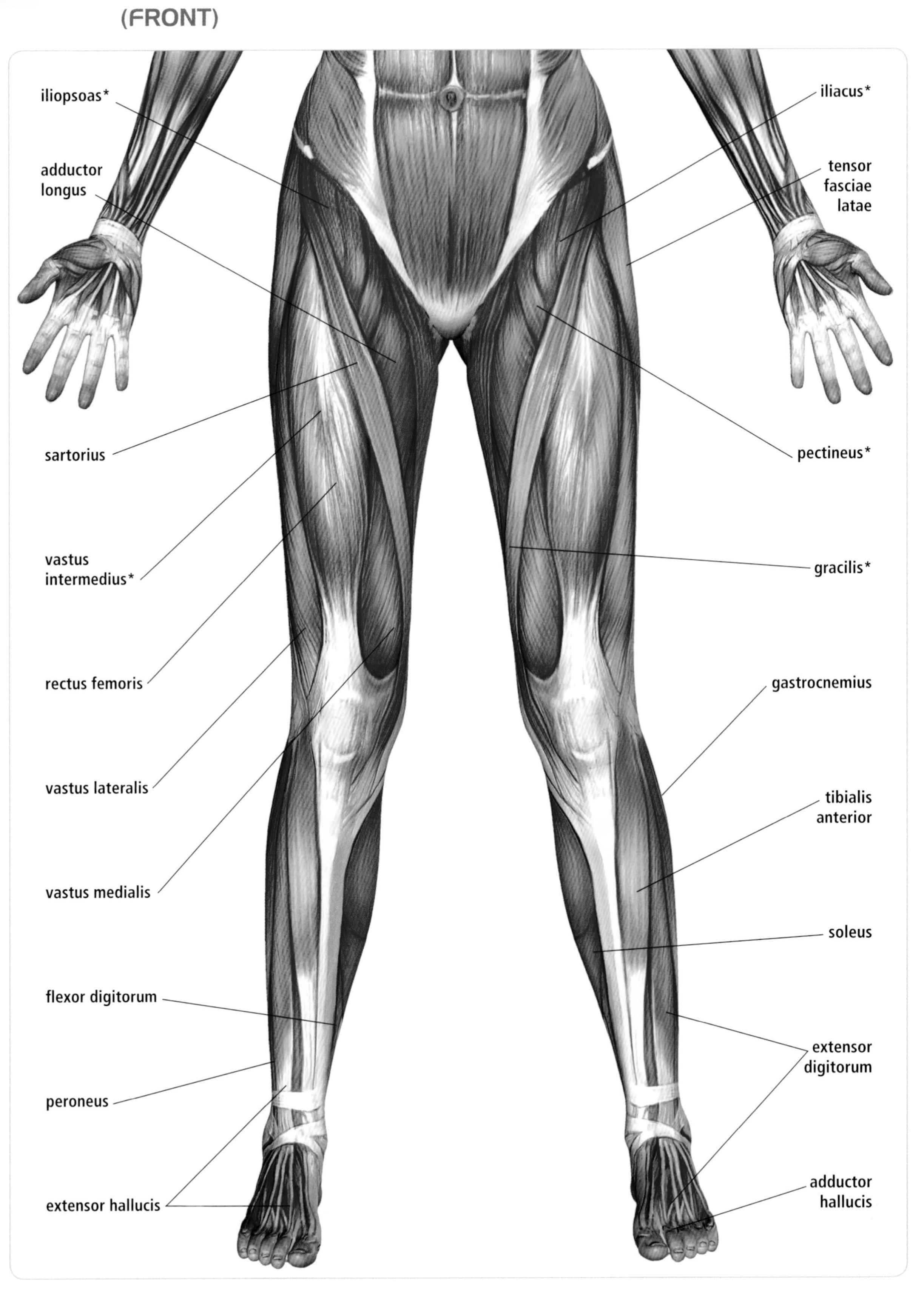

LOWER BODY
(BACK)

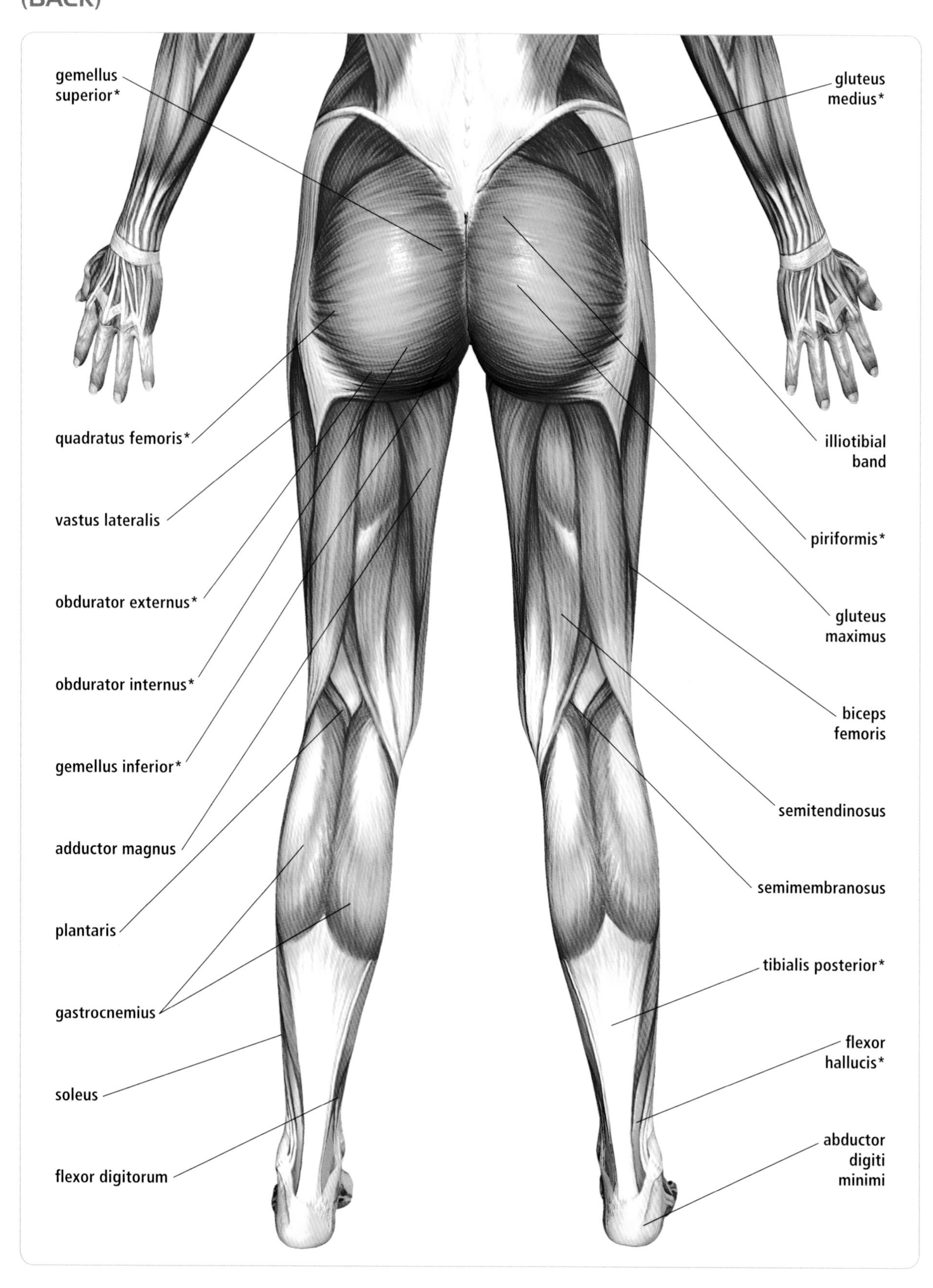

WARM-UP & COOL-DOWN

The warm-up and cool-down poses in yoga are the most important poses to ensure that you reap all the benefits of your routine. The first poses that you practice are meant to awaken your muscles, increase your heart rate, and release tension in your body, just as the last poses are meant to relax your muscles, lower your heart rate, and provide relief after an invigorating workout. Gentle stretching, especially after you exercise, is vital to injury prevention. In these poses, integrate your mind, body, and breath to find the focus that you need in your yoga practice. Poses such as the Easy Pose and Staff Pose form the foundation of most seated postures, while poses like the Knees-to-Chest Pose serve as counterposes to backbends.

EASY POSE
(SUKHASANA)

1 Sit on the floor with your legs extended in front of you.

2 Bend your knees, and cross your shins inward, sliding your left foot beneath your right knee and your right foot beneath your left knee, forming a gap between your feet and your groins. Relax your knees toward the floor.

3 Draw your sit bones to the floor, and lift up through your spine. Maintain a neutral position from your pelvis to your shoulders. Open your chest, and relax your shoulders.

4 Place the backs of your hands on your knees, forming an "O" with your thumb and index finger. Breathe slowly and evenly.

5 Hold for as long as you wish. Be sure to also practice the pose with your opposite leg in front.

DO IT RIGHT
- To help maintain neutrality in your pelvis, place the edge of a folded blanket beneath your sit bones.
- Relax the outsides of your feet on the floor.

AVOID
- Pulling your feet in toward your groins.
- Arching your lower back beyond its neutral spinal position.

PRONUNCIATION & MEANING
- Sukhasana (sook-AHS-anna)
- *sukh* = joy, comfort

LEVEL
- Beginner

BENEFITS
- Opens hips
- Strengthens spine
- Relieves stress

CONTRA-INDICATIONS & CAUTIONS
- Knee injury
- Hip injury

STAFF POSE
(DANDASANA)

DO IT RIGHT
- If your hamstrings are especially tight, place the edge of a folded blanket beneath your sit bones.
- Rotate your thighs slightly inward, pointing your knees up toward the ceiling.

AVOID
- Rolling back over your sit bones.

❶ Sit on the floor with your legs extended together in front of you. Draw your sit bones into the floor and away from your heels.

❷ Contract the muscles in your legs, pressing them against the floor. Place the palms of your hands on the floor beside your hips, and lift up through your spine. Flex your feet.

❸ Lift your chest, and gaze forward, tucking your chin slightly downward. Relax your shoulders, and pull your abdominals in toward your spine.

❹ Hold for 1 minute or more.

PRONUNCIATION & MEANING
- Dandasana (dan-DAHS-anna)
- *danda* = stick, staff

LEVEL
- Beginner

BENEFITS
- Strengthens spine
- Improves posture

CONTRA-INDICATIONS & CAUTIONS
- Lower-back injury

DOWNWARD-FACING DOG

(ADHO MUKHA SVANASANA)

1 Kneel on your hands and knees with your knees directly below your hips. Stretch your hands out slightly in front of your shoulders with your fingertips facing forward. They should be placed one shoulder-width apart.

DO IT RIGHT

- If your hamstrings and shoulders are especially tight, practice the pose with your knees slightly bent and your heels lifted from the floor.
- Contract your thighs to lengthen your spine further, and keep pressure off your shoulders.

AVOID

- Sinking your shoulders into your armpits, creating an arch in your back.
- Rounding your spine.

a

2 Exhale and press against the floor, keeping your elbows straight. Lift your sit bones up toward the ceiling and your knees away from the floor. Lengthen your hips away from your ribs to elongate your spine.

3 Press your heels toward the floor, and contract your thighs. Try to straighten your knees. Turn your thighs slightly inward, and broaden your chest and shoulders. Position your head in between your arms.

4 Hold for 30 seconds to 2 minutes.

b

PRONUNCIATION & MEANING

- Adho Mukha Svanasana (AH-doh MOO-kah shvah-NAHS-anna)
- *adho* = downward; *mukha* = face; *shvana* = dog

LEVEL

- Beginner

BENEFITS

- Stretches shoulders, hamstrings, and calves
- Strengthens arms and legs
- Relives stress and headaches

CONTRA-INDICATIONS & CAUTIONS

- Carpal tunnel syndrome

EXTENDED PUPPY POSE
(UTTANA SHISHOSANA)

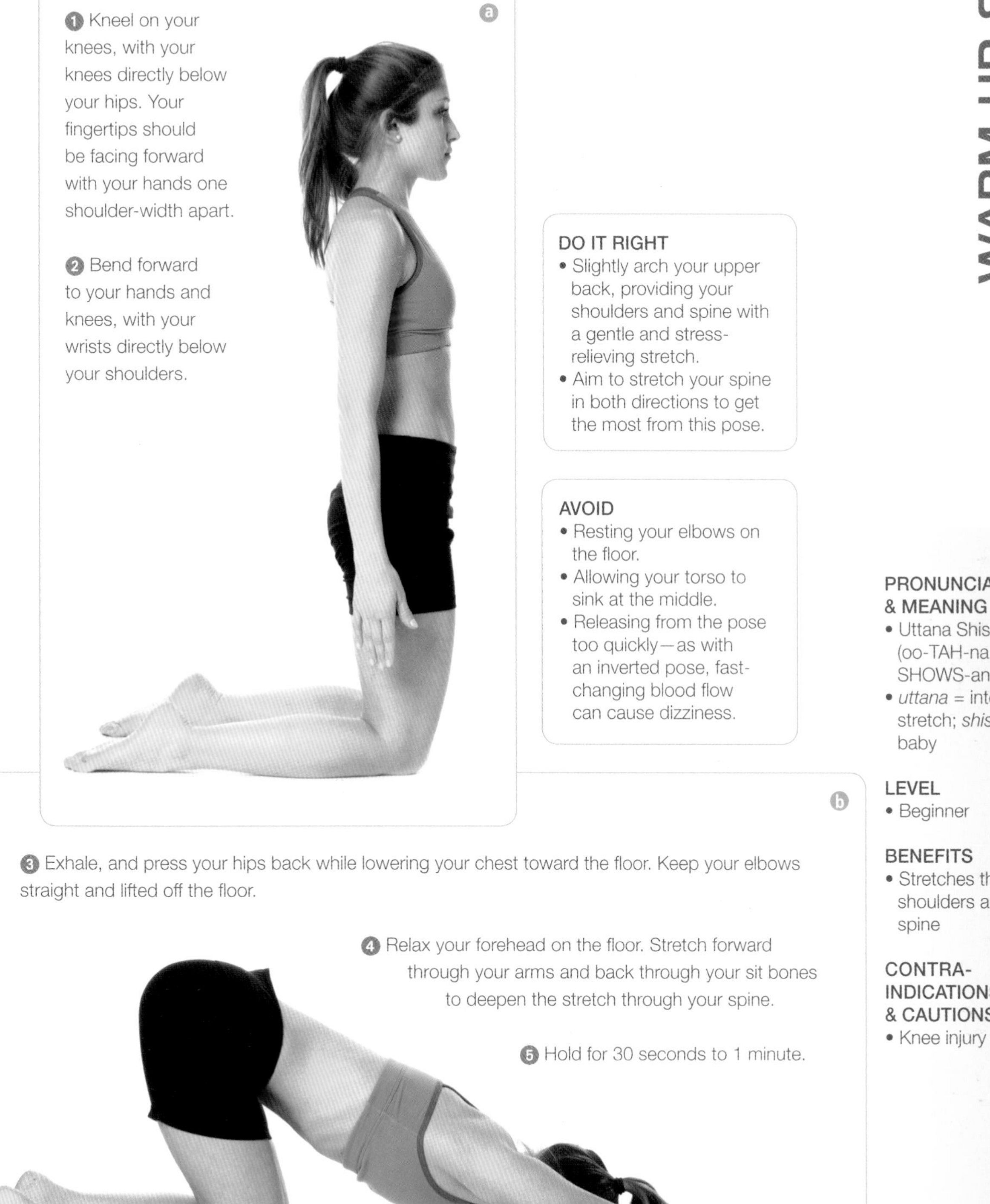

a

1. Kneel on your knees, with your knees directly below your hips. Your fingertips should be facing forward with your hands one shoulder-width apart.

2. Bend forward to your hands and knees, with your wrists directly below your shoulders.

b

3. Exhale, and press your hips back while lowering your chest toward the floor. Keep your elbows straight and lifted off the floor.

4. Relax your forehead on the floor. Stretch forward through your arms and back through your sit bones to deepen the stretch through your spine.

5. Hold for 30 seconds to 1 minute.

DO IT RIGHT
- Slightly arch your upper back, providing your shoulders and spine with a gentle and stress-relieving stretch.
- Aim to stretch your spine in both directions to get the most from this pose.

AVOID
- Resting your elbows on the floor.
- Allowing your torso to sink at the middle.
- Releasing from the pose too quickly—as with an inverted pose, fast-changing blood flow can cause dizziness.

PRONUNCIATION & MEANING
- Uttana Shishosana (oo-TAH-na shee-SHOWS-anna
- *uttana* = intense stretch; *shishu* = baby

LEVEL
- Beginner

BENEFITS
- Stretches the shoulders and spine

CONTRA-INDICATIONS & CAUTIONS
- Knee injury

CAT POSE TO COW POSE
(MARJARYASANA TO BITILASANA)

a

❶ Begin on your hands and knees, with your wrists directly below your shoulders and your knees directly below your hips. Your fingertips should be facing forward with your hands one shoulder-width apart. Gaze at the floor, keeping your head in a neutral position.

❷ Exhale, and round your spine up toward the ceiling, dropping your head. Draw your abdominals in toward your spine. Keep your hips lifted and your shoulders in the same position.

❸ Inhale, and uncurl your spine. Remain on your hands and knees.

DO IT RIGHT
- Draw your shoulders away from your neck.

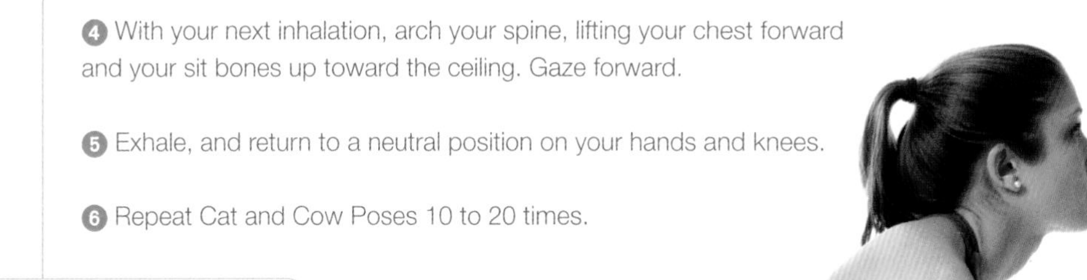

b

❹ With your next inhalation, arch your spine, lifting your chest forward and your sit bones up toward the ceiling. Gaze forward.

❺ Exhale, and return to a neutral position on your hands and knees.

❻ Repeat Cat and Cow Poses 10 to 20 times.

AVOID
- Arching primarily in your lower back.
- Tucking your chin to your chest in Cat Pose.
- Jutting your rib cage out in Cow Pose.

PRONUNCIATION & MEANING
- Marjaryasana (mar-jar-ee-AHS-anna) *marjari* = cat
- Bitilasana (bit-il-AHS-anna)
- There is no agreed-upon translation of the Sanskrit name for the Cow Pose.

LEVEL
- Beginner

BENEFITS
- Stretches shoulders, chest, abdominals, neck, and spine
- Relieves stress

CONTRA-INDICATIONS & CAUTIONS
- Knee injury

CHILD'S POSE
(BALASANA)

1. Kneel on the floor, with hips aligned over knees.

2. Bring your legs together so that your big toes are touching. Lower your body to rest your buttocks on your heels, and separate your knees about one hip-width apart.

3. Exhale, and lower your torso down to your inner thighs. Elongate your neck and your spine, stretching your tailbone down toward the floor.

4. Place the backs of your hands on the floor beside your feet. Allow your shoulders to relax toward the floor, widening them across your upper back. Place your forehead on the floor.

5. Hold 30 seconds to 3 minutes.

DO IT RIGHT
- Inhale into the back of your rib cage.
- Round your back to create a dome shape.

AVOID
- Compressing the back of your neck.

PRONUNCIATION & MEANING
- Balasana (bah-LAHS-anna)
- *bala* = child

LEVEL
- Beginner

BENEFITS
- Stretches spine, hips, thighs, and ankles
- Relieves stress

CONTRA-INDICATIONS & CAUTIONS
- Diarrhea
- Knee injury
- Pregnancy

KNEES-TO-CHEST POSE

(APANASANA)

a

1. Lie supine on the floor.

2. Exhale, and draw your knees toward your chest.

DO IT RIGHT
- If you can't grasp your elbows while hugging your knees, place your hands directly on your knees.
- Lengthen the back of your neck.

b

3. Wrap your arms around your knees, placing each hand on your opposite elbow. Lengthen the back of your neck away from your shoulders. With each exhalation, gently pull your knees closer toward your chest, and flatten your back and shoulders on the floor.

4. Hold for 30 seconds to 1 minute.

AVOID
- Tensing your back or leg muscles.

PRONUNCIATION & MEANING
- Apanasana (ap-AN-ahs-anna)
- *apana* = waste-eliminating downward breath

LEVEL
- Beginner

BENEFITS
- Stretches lower back and hips
- Stimulates digestion

CONTRA-INDICATIONS & CAUTIONS
- Knee injury
- Pregnancy

CORPSE POSE
(SAVASANA)

1. Sit on the floor on your buttocks with your knees bent. Lift your hips, and place your tailbone slightly closer to your heels. Elongate your lower back away from your tailbone before allowing your back to relax to the floor.

2. Straighten your legs one at a time. Allow your legs to fall open, separated the same distance from the center of your body. The feet should be turned out equally.

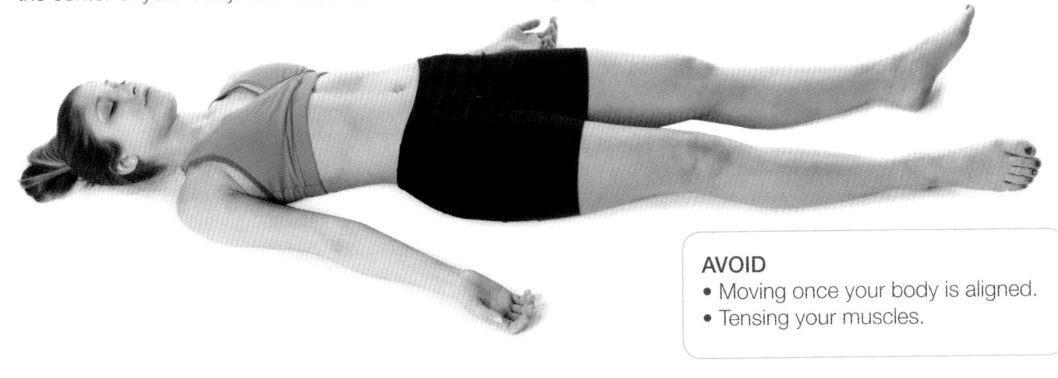

AVOID
- Moving once your body is aligned.
- Tensing your muscles.

3. Relax your arms on the floor by your sides, leaving a space between your torso and your arms. Spread your shoulder blades and your collarbones, and turn your arms out so that your palms face up.

4. Lengthen your neck away from your shoulders, and try to release it comfortably toward the floor. Close your eyes. Breathe smoothly. Focus on your body alignment and your breath.

5. Relax every part of your body, starting with your toes and ending with your head. Feel each part sinking into the floor. Relax the muscles in your face and calm your brain.

6. Hold for 5 to 10 minutes. Gently come out of the pose by bending your knees to your chest and rolling over to one side. Bring your head up last.

[alternate view]

DO IT RIGHT
- End your yoga practice with Corpse Pose.
- Pay attention to the alignment of your head, making sure that it is pulled away from your shoulders and does not tilt to one side.
- Practice with your knees bent and feet flat on the floor.

PRONUNCIATION & MEANING
- Savasana (shah-VAHS-anna)
- *sava* = corpse

LEVEL
- Beginner

BENEFITS
- Calms the brain
- Relieves stress
- Relaxes the body

CONTRA-INDICATIONS & CAUTIONS
- Back injury

STANDING POSES

Standing poses, generally practiced at the beginning of a yoga workout, build awareness of the fundamental movements of yoga. They energize your body, develop stamina, and revitalize your legs. Because they require strength, flexibility, and balance, standing poses give insight to what areas of your body are weak or unstable. When practicing the poses in this section, be aware of your body's alignment while you find a graceful balance. It is important that you ground your feet firmly and maintain good posture.

With such a wide range of movements performed in any standing series, you will stretch and increase mobility throughout your entire body. These poses will strengthen your arms, shoulders, torso, pelvis, legs, and feet. Your pelvis is the link between your torso and legs, and learning to stabilize your pelvis is key to mastering standing balances. This prepares you for other asanas, such as sitting poses.

MOUNTAIN POSE

(TADASANA)

1. Stand with your feet together, with both heels and toes touching.

2. Keeping your back straight and both arms pressed slightly against your sides, face your palms outward.

3. Lift all your toes and let them fan out, and then gently drop them down to create a wide, solid base.

4. Rock from side to side until you gradually bring your weight evenly onto all four corners of both feet.

5. While balancing your weight evenly on both feet, slightly contract the muscles in your knees and thighs, rotating both thighs inward to create a widening of the sit bones. Tuck your tailbone in between the sit bones.

6. Tighten your abdominals, drawing them in slightly, maintaining a firm posture.

7. Widen your collarbones, making sure that your shoulders are parallel to your pelvis.

8. Lengthen your neck, so that the crown of your head rises toward the ceiling, and your shoulder blades slide down your back.

9. Hold for 30 seconds to 1 minute.

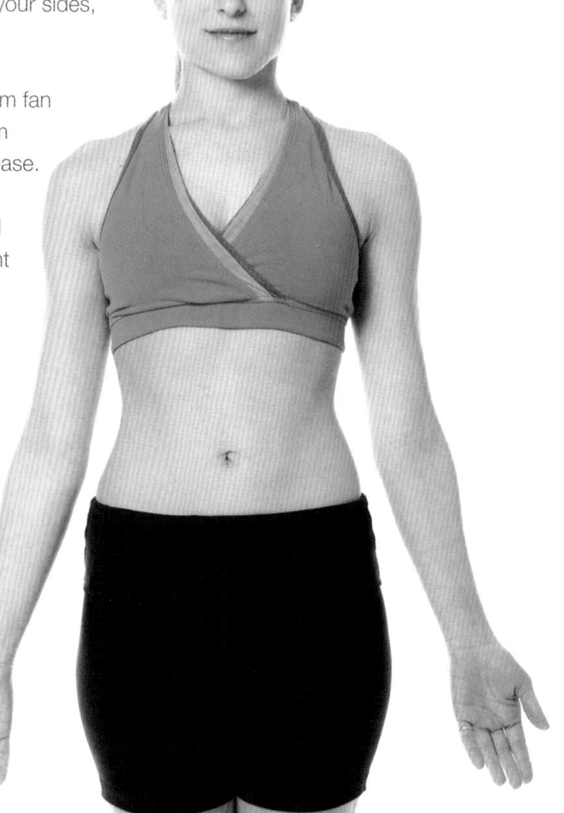

DO IT RIGHT

- If your ankles are knocking together uncomfortably, separate the heels slightly.
- If you are a beginner, practice the pose with your back to the wall to feel the alignment.

AVOID

- Slumping your back.
- Drooping your shoulders.

PRONUNCIATION & MEANING

- Tadasana (tah-DAHS-anna)
- *tada* = mountain

LEVEL

- Beginner

BENEFITS

- Improves posture
- Strengthens thighs

CONTRA-INDICATIONS & CAUTIONS

- Headache
- Insomnia
- Low blood pressure

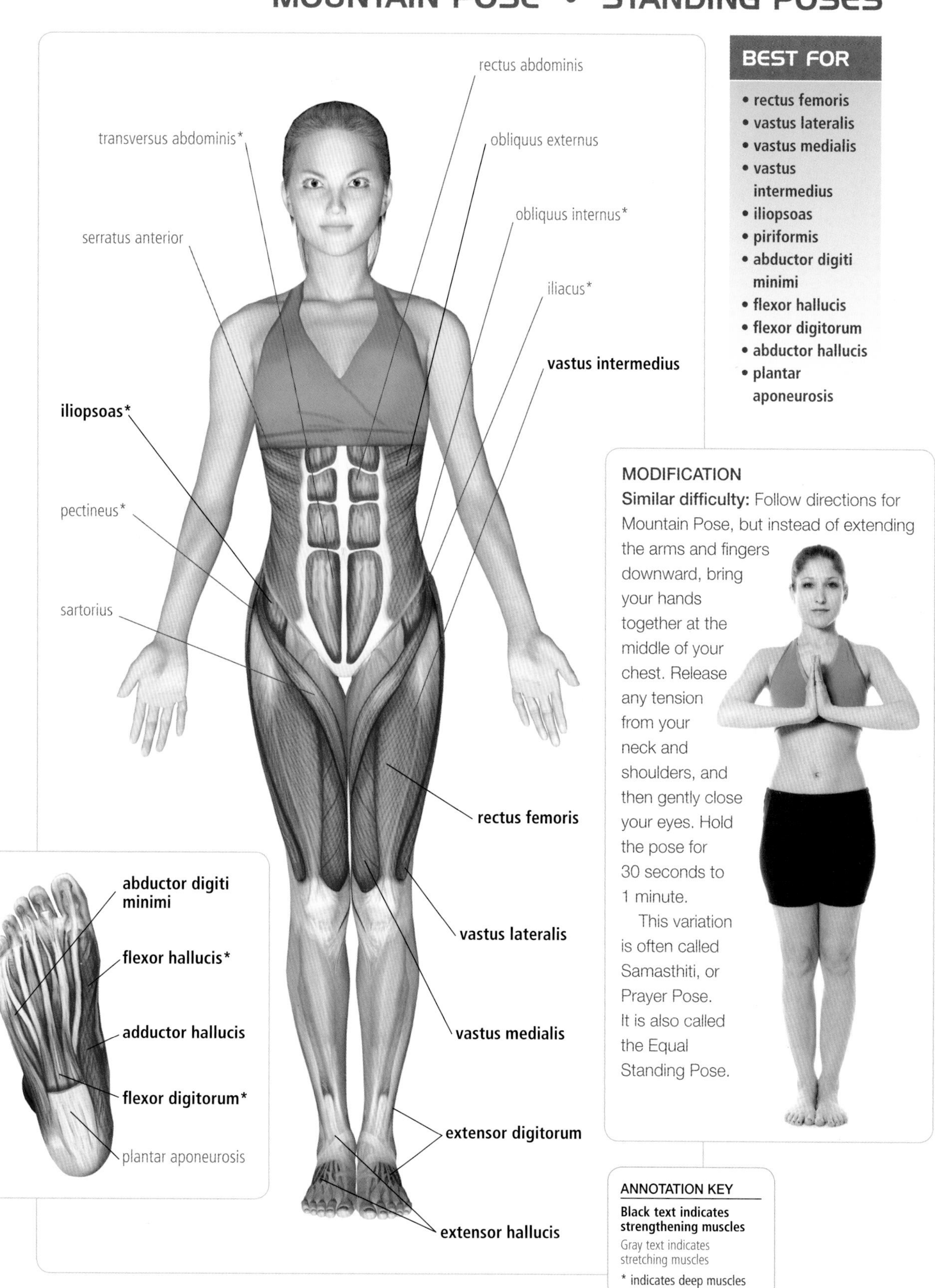

BEST FOR

- rectus femoris
- vastus lateralis
- vastus medialis
- vastus intermedius
- iliopsoas
- piriformis
- abductor digiti minimi
- flexor hallucis
- flexor digitorum
- abductor hallucis
- plantar aponeurosis

MODIFICATION

Similar difficulty: Follow directions for Mountain Pose, but instead of extending the arms and fingers downward, bring your hands together at the middle of your chest. Release any tension from your neck and shoulders, and then gently close your eyes. Hold the pose for 30 seconds to 1 minute.

This variation is often called Samasthiti, or Prayer Pose. It is also called the Equal Standing Pose.

ANNOTATION KEY

Black text indicates strengthening muscles

Gray text indicates stretching muscles

* indicates deep muscles

GARLAND POSE
(MALASANA)

1 Stand in Mountain Pose (Tadasana, see page 32), with feet shoulder-width apart and your pelvis, head, and chest aligned.

2 Keeping your heels on the floor, extend your arms straight out in front of you. Bend your knees, folding your body forward and down by dropping your pelvis.

3 Slightly separate your thighs wider than your torso. Exhale, and lean your body forward, fitting it snugly in the space between your thighs.

4 Press your elbows against the back of your knees, and join your palms together, as if in prayer, and then press your knees into your elbows.

5 Hold for 30 seconds to 1 minute. Exhale, and straighten your knees, slowly standing up.

AVOID
- Leaning forward.
- Drooping your shoulders.

DO IT RIGHT
- If your heels come up as you reach the squatting position, place a folded blanket under them, and squat again.
- If squatting is difficult, you can get a similar stretch by sitting on the front edge of a chair seat, with your thighs forming a right angle to your torso. Place your heels on the floor slightly in front of your knees, and lean your torso forward between the thighs.

PRONUNCIATION & MEANING
- Malasana (ma-LAHS-anna)
- *mala* = garland
- Also called Wide Squat or Frog Pose

LEVEL
- Beginner

BENEFITS
- Stretches ankles, groins, lower legs, and back torso
- Tones pelvic-floor muscles
- Tones abdominals

CONTRA-INDICATIONS & CAUTIONS
- Headache
- Insomnia
- Low blood pressure

BEST FOR

- quadratus lumborum*
- quadratus femoris
- transversus abdominis
- biceps femoris
- sartorius
- vastus intermedius
- vastus medialis
- vastus lateralis
- semitendonosus
- semimembranosus

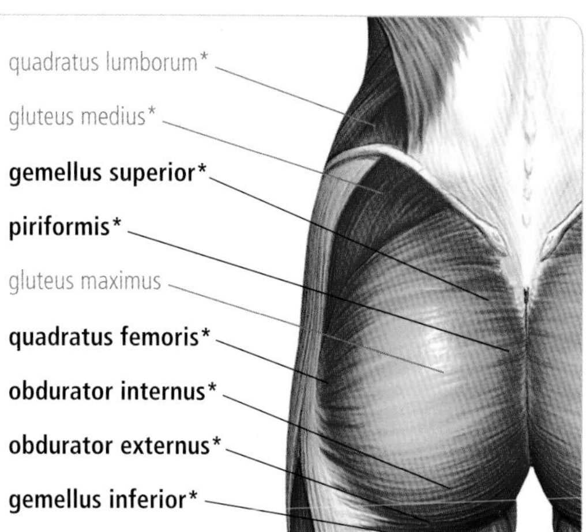

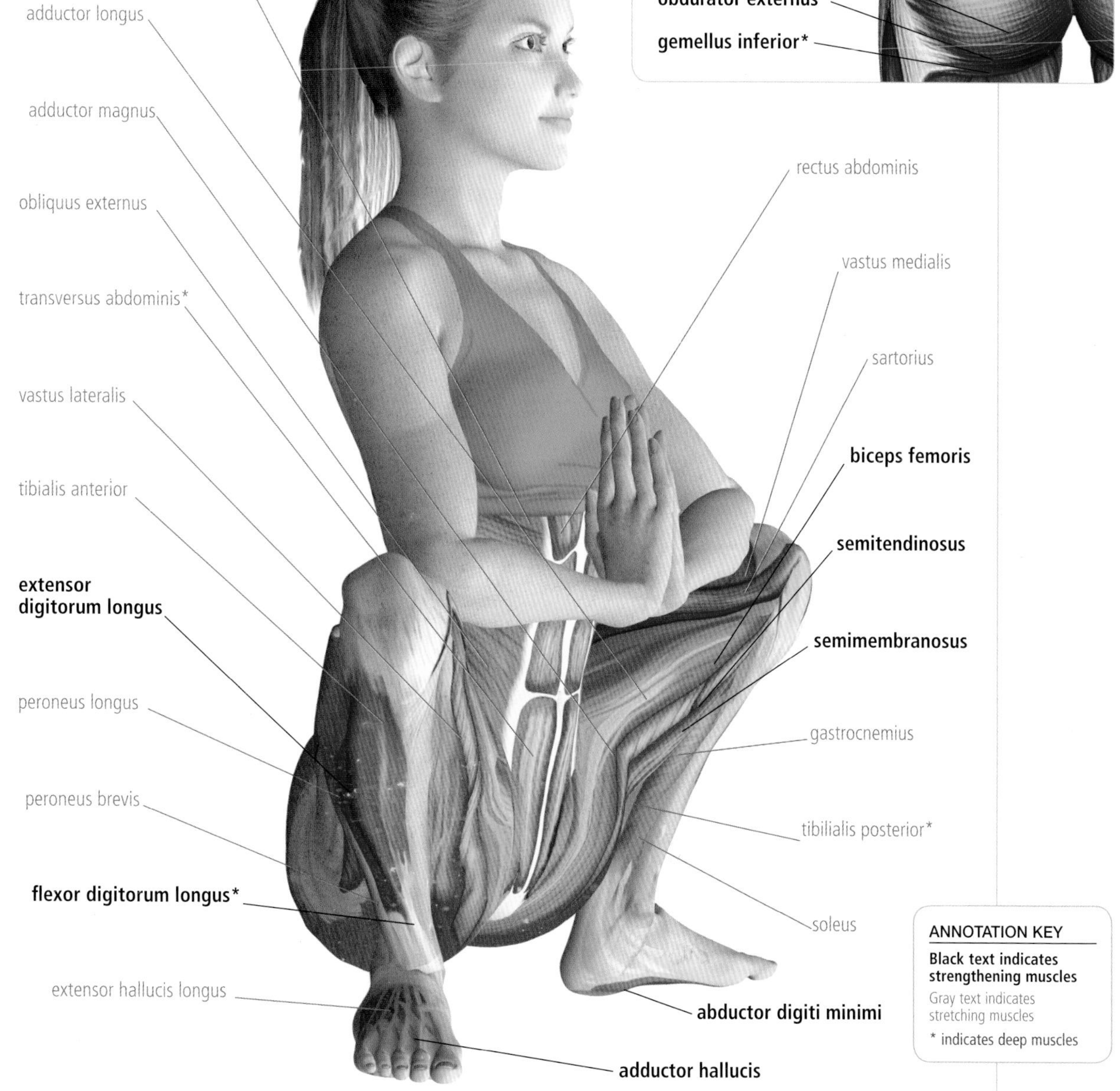

ANNOTATION KEY

Black text indicates strengthening muscles

Gray text indicates stretching muscles

* indicates deep muscles

UPWARD SALUTE
(URDHVA HASTASANA)

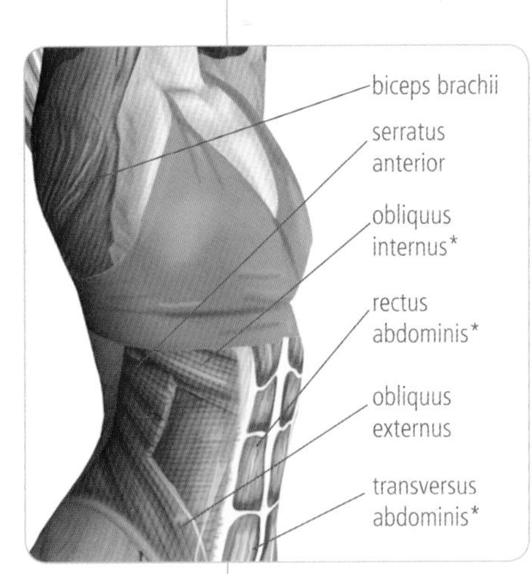

PRONUNCIATION & MEANING
- Urdhva Hastasana (oord-vah hahs-TAHS-anna)
- *urdhva* = raised (or upward); *hasta* = hand
- Also called Raised Hand Pose

LEVEL
- Beginner

BENEFITS
- Fights fatigue
- Relieves indigestion
- Alleviates back aches
- Stretches abdominals
- Stretches shoulders and armpits
- Relieves mild anxiety

CONTRA-INDICATIONS & CAUTIONS
- Shoulder injury
- Neck injury

1. Stand in Mountain Pose (Tadasana, see page 32), with your feet shoulder-width apart and your pelvis, head, and chest aligned. Turn your palms inward.

2. Keeping your arms parallel and palms facing each other, inhale, and sweep your arms out in front of you to the height of your shoulders, and then alongside your ears, raising them upward toward the ceiling.

3. Spread your shoulder blades and draw your chin in slightly, as you gently tip your head back. Gaze at your thumbs.

4. Hold for 30 seconds to 1 minute.

5. Exhale, pulling your hands down with your palms together. As your hands lower toward your face, gently drop your head until it returns to a neutral position.

DO IT RIGHT
- Keep your shoulders aligned directly over your hips and your hips over your heels.
- Keep back ribs broad.
- Broaden the top of your shoulder blades.
- Move your armpits down while lifting the arms upward.

AVOID
- Jutting your rib cage out of your chest.

BEST FOR
- **obliquus externus**
- **obliquus internus**
- **transversus abdominis**
- **latissimus dorsi**
- **teres major**
- **infraspinatus**

ANNOTATION KEY
Black text indicates strengthening muscles
Gray text indicates stretching muscles
* indicates deep muscles

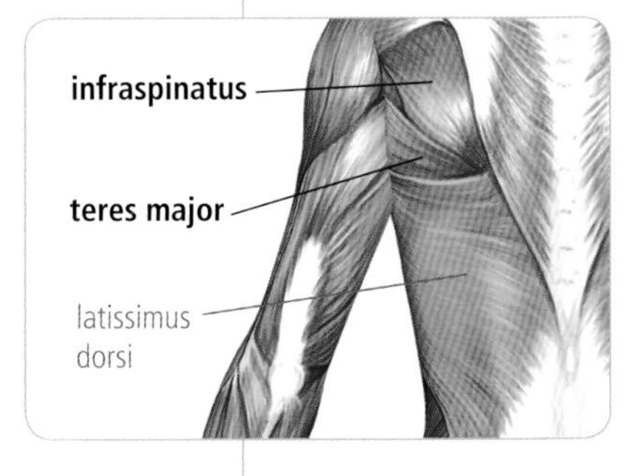

AWKWARD POSE
(UTKATASANA)

① Stand in Mountain Pose (Tadasana, see page 32). Inhale, and raise both your hands over your head, keeping your arms straight and lengthening your spine. You may clasp your hands together or keep them shoulder-width apart.

② Exhale, and bend your knees. Bend your upper body forward so that it is at a 45-degree angle to the floor, keeping your lower back straight. Relax your calf muscles, allowing the weight of your upper body to sink into your pelvis. Transfer your weight to your heels.

③ Hold for 30 seconds to 1 minute.

④ Inhale, and straighten your knees, lifting strongly through your arms. Exhale, release your arms to your side, and return to Mountain Pose.

BEST FOR

- erector spinae
- extensor digitorum
- triceps brachii
- deltoideus
- infraspinatus
- teres major
- gluteus medius
- biceps femoris
- semitendinosus
- semimembranosus
- soleus
- tibialis anterior
- rectus femoris
- vastus lateralis
- vastus medialis
- vastus intermedius

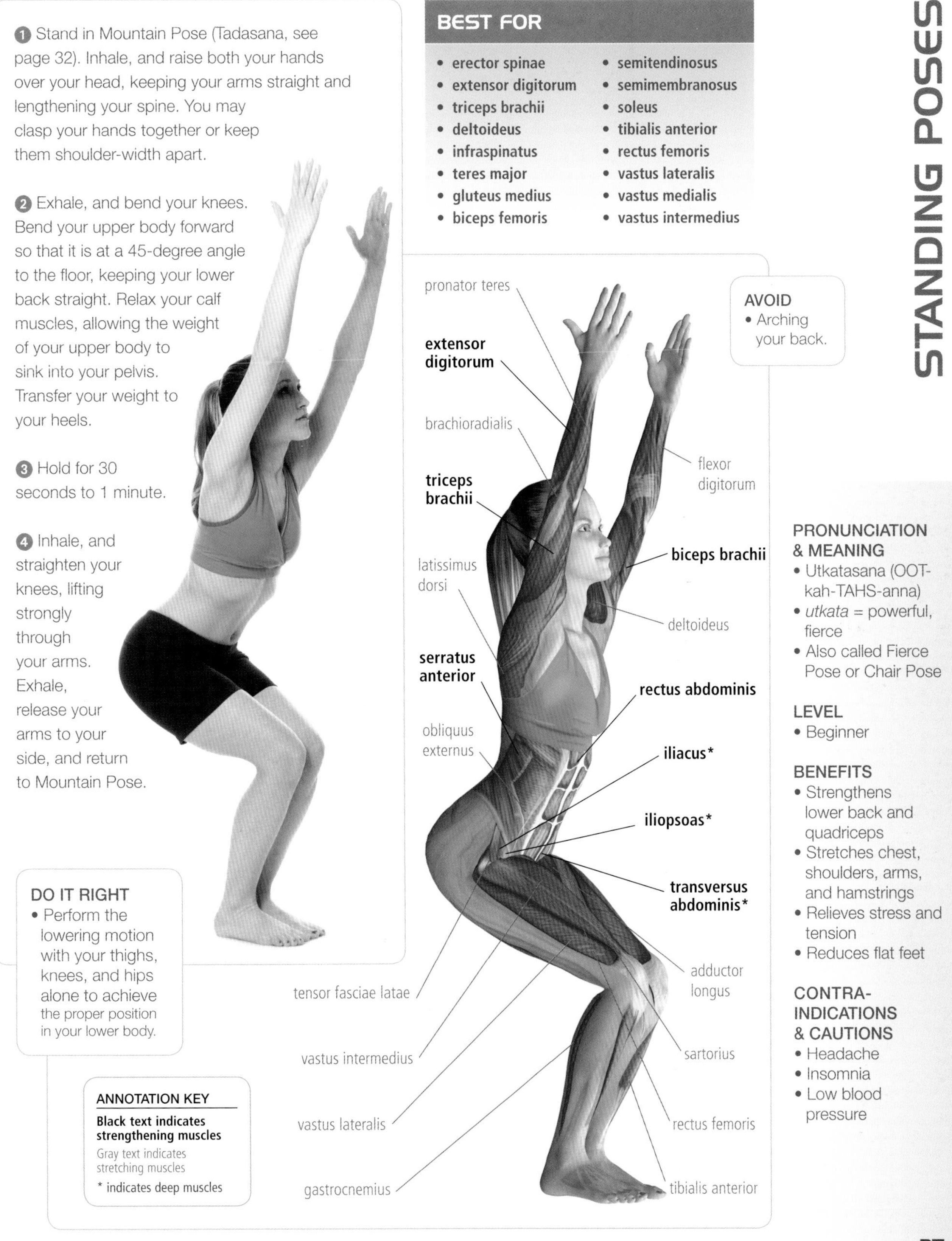

AVOID
- Arching your back.

DO IT RIGHT
- Perform the lowering motion with your thighs, knees, and hips alone to achieve the proper position in your lower body.

ANNOTATION KEY
Black text indicates strengthening muscles
Gray text indicates stretching muscles
* indicates deep muscles

PRONUNCIATION & MEANING
- Utkatasana (OOT-kah-TAHS-anna)
- *utkata* = powerful, fierce
- Also called Fierce Pose or Chair Pose

LEVEL
- Beginner

BENEFITS
- Strengthens lower back and quadriceps
- Stretches chest, shoulders, arms, and hamstrings
- Relieves stress and tension
- Reduces flat feet

CONTRA-INDICATIONS & CAUTIONS
- Headache
- Insomnia
- Low blood pressure

TREE POSE
(VRKSASANA)

❶ Stand in Prayer Pose (Samasthiti, see page 33). Shift your weight slightly onto your left foot, keeping your inner foot firmly grounded on the floor. Bend your right knee, and reach down with your right hand and grasp your right ankle.

❷ Draw your right foot up, and place the sole against your inner left thigh. Press your right heel into your inner left groin, while pointing your toes toward the floor. The center of your pelvis should be directly over the left foot.

❸ Rest your hands on the top rim of your pelvis. Make sure the pelvis is in a neutral position, with the top rim parallel to the floor.

❹ Lengthen your tailbone toward the floor. Firmly press the sole of your right foot against your inner thigh while resisting with your outer left leg. Press your hands together, and gaze at a fixed point in front of you on the floor about 4 feet to 5 feet away.

❺ Hold for 30 seconds to 1 minute. Exhale, and step back into Prayer Pose. Repeat with your opposite leg standing.

AVOID
- Jutting out your hip—keep both hips squared forward.

DO IT RIGHT
- If you are a beginner, brace your back against a wall to steady yourself.
- To keep your raised foot from sliding, place a folded sticky mat between your sole and inner thigh.

PRONUNCIATION & MEANING
- Vrksasana (vrik-SHAHS-anna)
- *vrksa* = tree

LEVEL
- Beginner

BENEFITS
- Strengthens thighs, calves, ankles, and spine
- Stretches groins, inner thighs, chest, and shoulders
- Improves sense of balance
- Relieves sciatica
- Reduces flat feet

CONTRA-INDICATIONS & CAUTIONS
- Headache
- Insomnia
- High or low blood pressure

BEST FOR

- iliacus
- iliopsoas
- gluteus maximus
- gluteus medius
- piriformis
- adductor magnus
- obdurator internus
- obdurator externus
- tensor fasciae latae
- rectus femoris

quadratus lumborum*
gluteus medius*
piriformis*
gluteus maximus
quadratus femoris*
obdurator internus*
obdurator externus*

ANNOTATION KEY
Black text indicates strengthening muscles
Gray text indicates stretching muscles
* indicates deep muscles

MODIFICATION
More difficult:
Follow steps 1 through 4, and then raise both arms over the head, keeping the elbows straight. Join the palms together. Hold for 30 seconds to 1 minute. Lower the arms and right leg and return to Prayer Pose. Pause for a few moments, and repeat on the opposite leg.

obliquus internus*
rectus abdominis
obliquus externus
tensor fasciae latae
transversus abdominis
rectus femoris
vastus medialis
gastrocnemius
tibialis anterior
soleus
iliopsoas*
iliacus*
pectineus*
adductor longus
adductor longus

EAGLE POSE
(GARUDASANA)

❶ Stand in Mountain Pose (Tadasana, see page 32), with your feet shoulder-width apart and your pelvis, head, and chest aligned.

❷ Shift your weight to your right leg, and then bend your knees slightly. Lift your left foot as you balance on your right foot, and cross your left thigh over the right.

❸ Point your left toes toward the floor, press your foot back, and then hook the top of your foot behind your lower right calf. Maintain your balance on your right foot.

❹ Inhale, and stretch your arms straight forward, keeping them parallel to the floor, and spread your scapulas wide across your back. Cross the arms in front of your torso so that your right arm is above the left, and then bend your elbows. Bring your right elbow into the crook of the left, and raise your forearms so that they are perpendicular to the floor. The backs of your hands should be facing each other.

❺ Press your right hand to the right and your left hand to the left, so that your palms face each other. Your right-hand thumb should pass in front of the little finger of your left hand. Press your palms together, lift your elbows up, and stretch your fingers toward the ceiling.

❻ Hold for 15 to 60 seconds.

❼ Slowly unwind your legs and arms, and return to Mountain Pose. Repeat with your arms and legs reversed.

AVOID
- Shifting your hips. Keep your hips squared to the front of your mat.

PRONUNCIATION & MEANING
- Garudasana (gah-rue-DAHS-anna)
- *garuda* = eagle, or the name of a mythic king of birds

LEVEL
- Beginner

BENEFITS
- Strengthens ankles and calves
- Stretches ankles, calves, thighs, hips, shoulders, and upper back
- Improves concentration
- Improves sense of balance

CONTRA-INDICATIONS & CAUTIONS
- Arm injury
- Hip injury
- Knee injury

DO IT RIGHT

- If you find it difficult to wrap your arms around each other until your palms touch, stretch your arms straight forward, parallel to the floor, while holding onto the ends of a strap.
- If you find it difficult to maintain your balance as you hook the foot of your raised leg behind the calf of your standing leg, press the big toe of your raised-leg foot against the floor.

MODIFICATION

More difficult: Follow steps 1 through 5. Sink down on your right foot, bending both knees as you move down. Bend forward from your hips, with your head facing your crossed arms. Hold for 15 to 60 seconds.

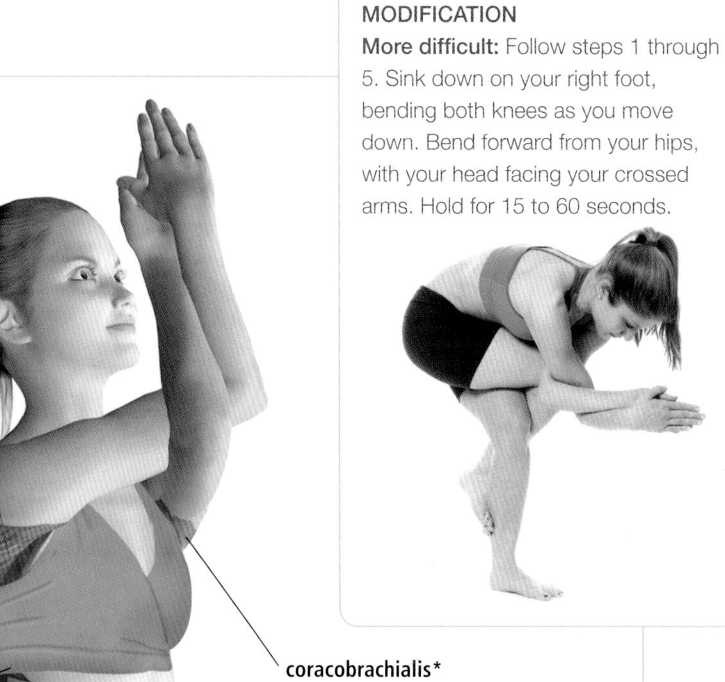

BEST FOR

- **trapezius**
- **infraspinatus**
- **teres major**
- **teres minor**
- **latissimus dorsi**
- **gluteus medius**
- **adductor magnus**
- **quadratus lumborum**
- **serratus anterior**

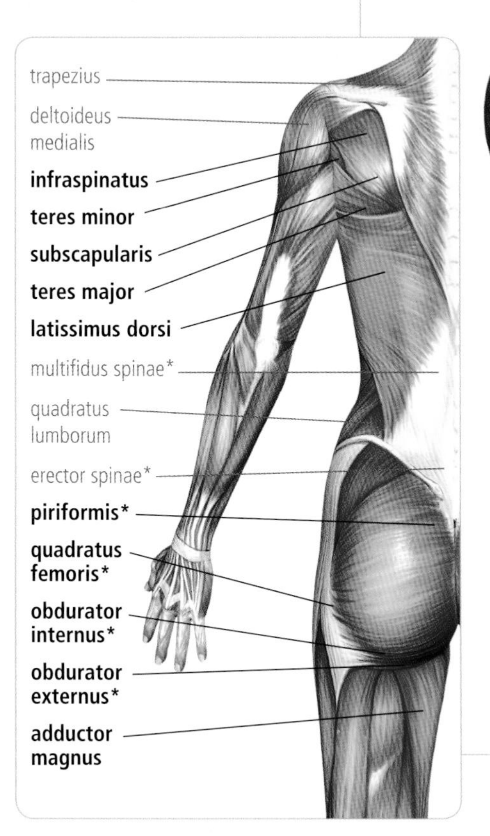

ANNOTATION KEY

Black text indicates strengthening muscles

Gray text indicates stretching muscles

* indicates deep muscles

TRIANGLE POSE
(TRIKONASANA)

a

1 Stand in Mountain Pose (Tadasana, see page 32), with your pelvis, head, and chest aligned.

2 Separate your feet slightly farther than the width of your shoulders.

3 Inhale, and raise both arms straight out to the side, keeping them parallel to the floor with your palms facing down.

4 Exhale slowly, and without bending your knees, pivot on your heels to turn your right foot all the way to the right and your left foot slightly toward the right, keeping your heels in line with each other.

5 Drop your torso as far as is comfortable to the right side, keeping your arms parallel to the floor.

AVOID
- Twisting your hips.

b

6 Once the torso is fully extended to the right, drop your right arm so that your right hand rests on your shin or on the front of your ankle. At the same time, extend your left arm straight up toward the ceiling. Gently twist your spine and torso counterclockwise, using your extended arms as a lever, while your spinal axis remains parallel to the ground. Extend your arms apart from each other in opposite directions.

c

7 Turn your head to gaze at your left thumb, slightly intensifying the twist in your spine. Hold for 30 seconds to 1 minute.

8 Inhale, and return to a standing position with the arms outstretched, strongly pressing the back heel into the floor. Reverse the feet, and repeat on the other side.

PRONUNCIATION & MEANING
- Trikonasana (trik-cone-AHS-anna)
- *trikona* = three angles, or triangle

LEVEL
- Beginner

BENEFITS
- Stretches thighs, knees, ankles, hips, groins, hamstrings, calves, shoulders, chest, and spine
- Relieves stress
- Stimulates digestion
- Relieves the symptoms of menopause
- Relieves backache

CONTRA-INDICATIONS & CAUTIONS
- Diarrhea
- Headache
- High or low blood pressure
- Neck issues

BEST FOR

- gluteus medius
- tensor fasciae latae
- sartorius
- piriformis
- serratus anterior
- obliquus externus
- latissimus dorsi

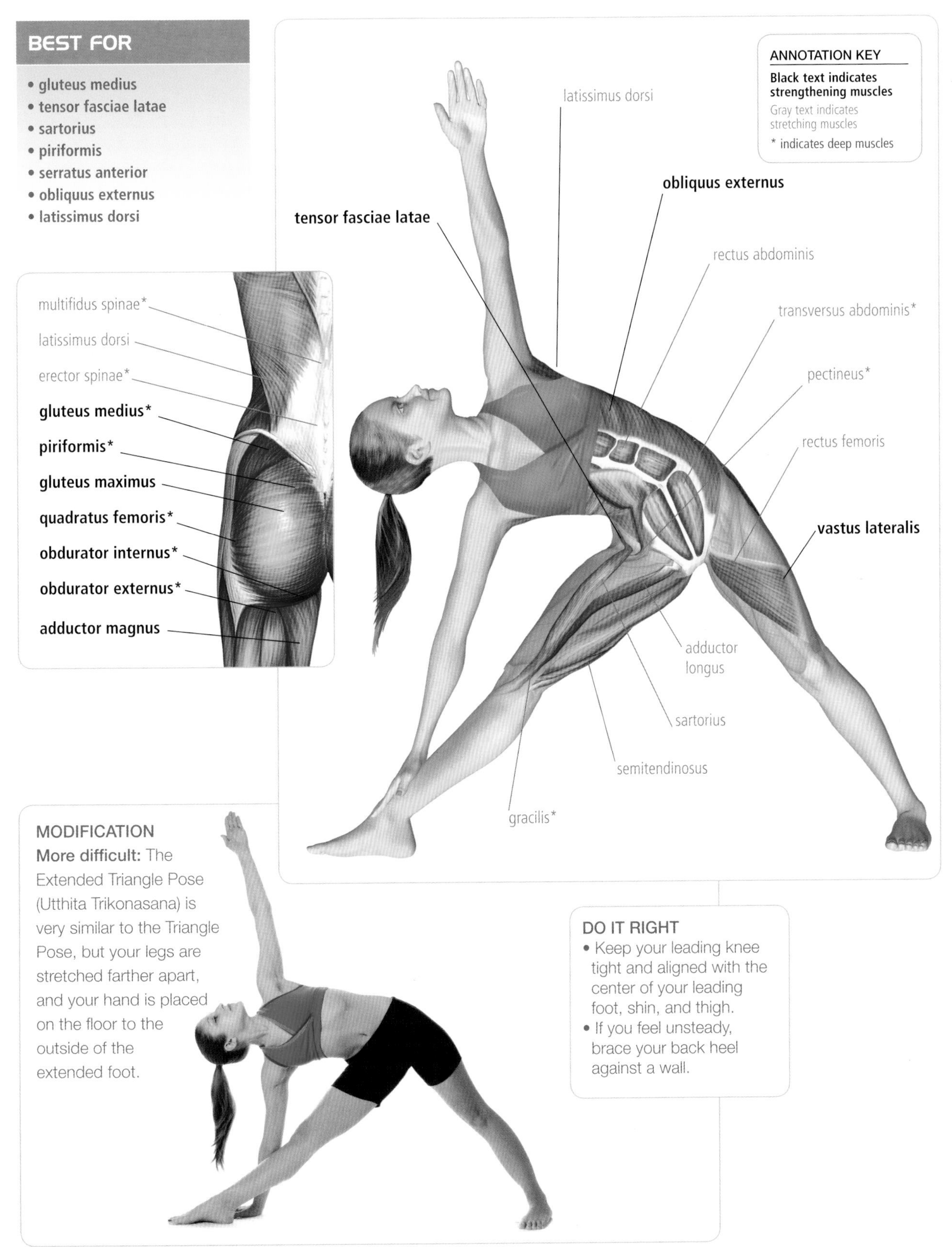

MODIFICATION

More difficult: The Extended Triangle Pose (Utthita Trikonasana) is very similar to the Triangle Pose, but your legs are stretched farther apart, and your hand is placed on the floor to the outside of the extended foot.

DO IT RIGHT

- Keep your leading knee tight and aligned with the center of your leading foot, shin, and thigh.
- If you feel unsteady, brace your back heel against a wall.

REVOLVED TRIANGLE POSE

(PARIVRTTA TRIKONASANA)

❶ Stand in Mountain Pose (Tadasana, see page 32). Exhale, and step or lightly jump your feet 3.5 to 4 feet apart.

❷ Raise your arms parallel to the floor and reach them out to the sides, with your shoulder blades wide, palms down. Turn your left foot in 45 to 60 degrees to the right and your right foot out 90 degrees to the right. Align your heels with each other, contract your thigh muscles, and turn your right thigh outward, so that the center of your right kneecap is in line with the center of your right ankle.

❸ Exhale, and turn your torso to the right, squaring your hip points with the front edge of your mat. As you bring your left hip around to the right, firmly ground your left heel. Inhale.

❹ Exhale, and turn your torso farther to the right and lean forward over your front leg. Reach your left hand down, either to the floor or on either side of your right foot. Allow your left hip to drop slightly toward the floor.

❺ Turn your head to gaze up at your top thumb. Widen the space between your shoulder blades, pressing your arms away from your torso. Shift most of your weight to your back heel and front hand.

❻ Hold for 30 seconds to 1 minute. Exhale, release the twist, and bring your torso back to upright with an inhalation. Repeat for the same length of time with the legs reversed, twisting to the left.

DO IT RIGHT

- Keep your hips level and parallel to the floor.
- If your are a beginner, rather than gaze upward, keep your head in a neutral position, looking straight forward, or turn your head to look at the floor.
- If you feel your hip slip out to the side and lift up toward your shoulder, press your outer right thigh actively to the left and release your right hip away from your right shoulder.

AVOID

- Shifting your hips to either side.

PRONUNCIATION & MEANING

- Parivrtta (par-ee-vrit-tah trik-cone-AHS-anna)
- *parivrtta* = to turn around, revolve; *trikona* = three angles, triangle

LEVEL

- Intermediate

BENEFITS

- Strengthen legs
- Stretches groins, hamstrings, and hips
- Opens chest and shoulders
- Cleanses internal organs

CONTRA-INDICATIONS & CAUTIONS

- Low blood pressure
- Migraine
- Diarrhea
- Insomnia

BEST FOR

- rectus femoris
- biceps femoris
- gluteus maximus
- gluteus medius
- obliquus internus
- obliquus externus
- latissimus dorsi
- erector spinae

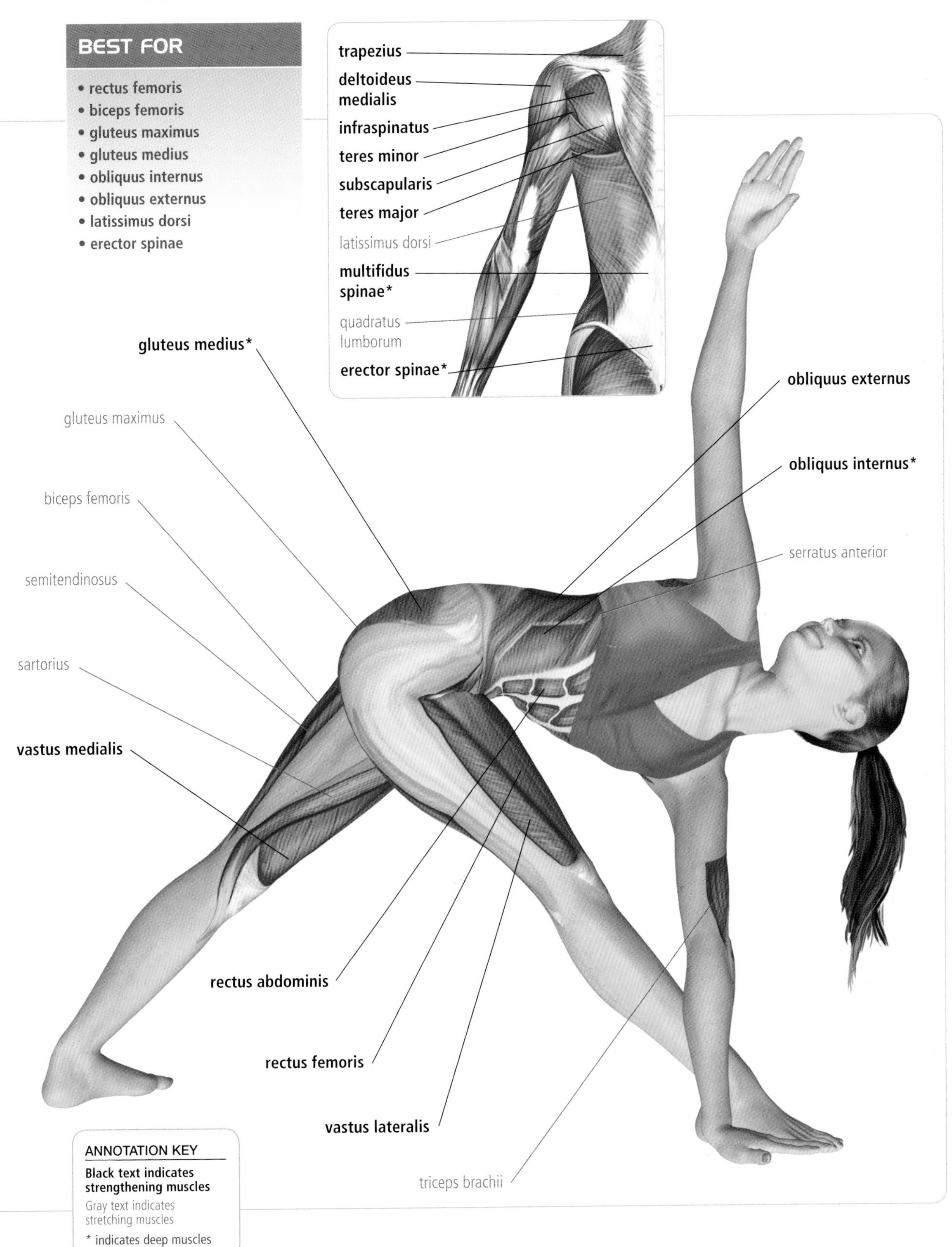

ANNOTATION KEY
Black text indicates strengthening muscles
Gray text indicates stretching muscles
* indicates deep muscles

HALF MOON POSE
(ARDHA CHANDRASANA)

a

❶ Stand in Triangle Pose (Trikonasana, see pages 42–43) to the right side, and rest your left hand on your left hip.

b

❷ Inhale, and with your right knee still bent, slide your left foot forward about 6 to 12 inches. At the same time, reach your right hand forward, beyond the little-toe side of your right foot, at least 12 inches.

c

DO IT RIGHT
- Actively elongate your raised leg from your hip through your heel, keeping it strong.

❸ Exhale, press your right hand and right heel firmly into the floor, and straighten your right leg, simultaneously lifting your left leg parallel to the floor.

❹ Rotate your upper torso to the left, while moving your left hip slightly forward. Most of your weight should rest on your standing leg. Press your right hand lightly against the floor, using it to maintain your balance.

❺ Hold for 30 seconds to 1 minute. Exhale, lower your raised leg to the floor, and return to Triangle Pose. Repeat on the other side, starting with your left leg bent.

PRONUNCIATION & MEANING
- Ardha Chandrasana (are-dah chan-DRAHS-anna)
- *ardha* = half; *candra* = moon, glittering, shining

LEVEL
- Intermediate

BENEFITS
- Strengthens spine, abdominals, ankles, thighs, and buttocks
- Stretches groins, hamstrings and calves, shoulders, chest, and spine
- Improves sense of balance
- Relieves stress
- Stimulates digestion

CONTRA-INDICATIONS & CAUTIONS
- Headache
- Diarrhea
- Low blood pressure

AVOID
- Locking your standing knee.
- Turning the kneecap of your standing leg inward—your kneecap should be aligned straight forward.

BEST FOR
- **latissimus dorsi**
- **obliquus internus**
- **obliquus externus**
- **serratus anterior**
- **transversus abdominis**
- **rectus abdominis**
- **vastus medialis**
- **biceps femoris**

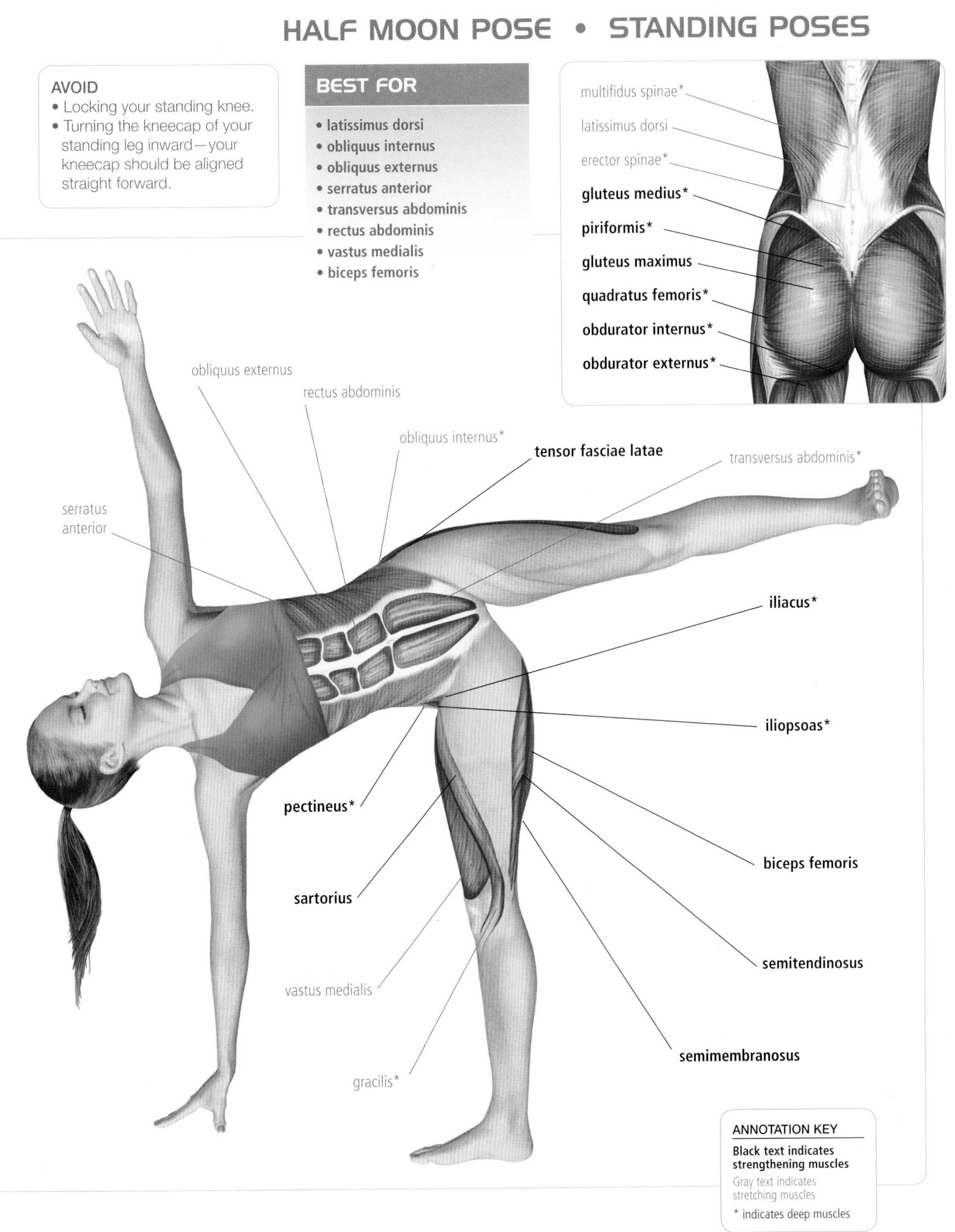

ANNOTATION KEY
Black text indicates strengthening muscles
Gray text indicates stretching muscles
* indicates deep muscles

EXTENDED HAND-TO-BIG-TOE POSE

(UTTHITA HASTA PADANGUSTHASANA)

BEST FOR

- rectus femoris
- vastus lateralis
- vastus medialis
- pronator teres
- flexor carpi radialis
- palmaris longus
- biceps femoris
- semitendinosus
- semimembranosus
- quadratus lumborum
- piriformis
- gemellus superior
- gemellus inferior
- tibialis anterior
- gracilis
- gluteus maximus

a

1 Stand in Mountain Pose (Tadasana, see page 32). Shift your weight onto your right foot. Firmly ground your right foot, pressing all corners of your foot and toes into the floor.

2 Square your hips facing forward, and raise your left leg toward your chest by bending your left knee. Grasp your left big toe with two fingers of your left hand curled around it. Rest your right hand on your right hip.

3 Exhale, and extend your left leg, straightening it while pulling your foot inward as your extended leg moves to come in line with your torso.

4 Gaze at a single spot on the floor about a body's length in front of you. Flex your foot so that your toes curl back toward you. Hold for about 30 seconds.

5 Exhale, and lower your foot to the floor. Repeat on the other side.

b

PRONUNCIATION & MEANING
- Utthita Hasta Padangusthasana (oo-TEET-uh HAWS-tuh POD-ung-goos-TAWS-uh-nuh)
- *utthita* = extended; *hasta* = hand; *pandangustha* = big toe

LEVEL
- Intermediate

BENEFITS
- Strengthens legs and ankles
- Stretches backs of the legs
- Improves sense of balance

CONTRA-INDICATIONS & CAUTIONS
- Ankle injury
- Lower-back injury

AVOID
- Moving the raised leg's hip up toward the lower ribs, so that your hips are no longer aligned.

DO IT RIGHT
- Keep your hips squared, facing forward—even when you raise your leg.
- Extend your torso, keeping as much space between your sternum and pubic bone as possible.

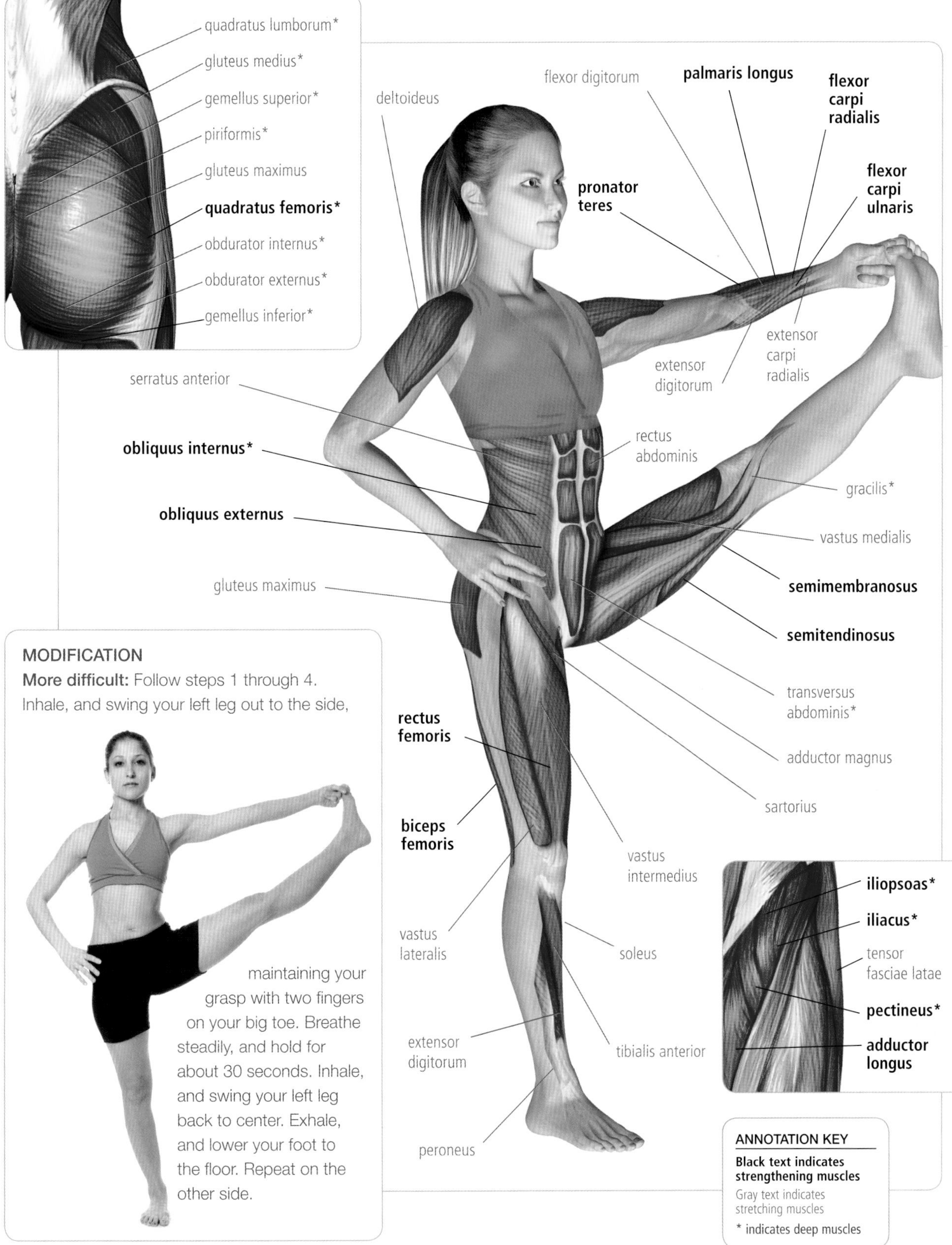

ANNOTATION KEY

Black text indicates strengthening muscles

Gray text indicates stretching muscles

* indicates deep muscles

MODIFICATION

More difficult: Follow steps 1 through 4. Inhale, and swing your left leg out to the side, maintaining your grasp with two fingers on your big toe. Breathe steadily, and hold for about 30 seconds. Inhale, and swing your left leg back to center. Exhale, and lower your foot to the floor. Repeat on the other side.

LOW LUNGE POSE
(ANJANEYASANA)

a

1 Stand in Downward-Facing Dog Pose (Adho Mukha Svanasana, see page 24). Exhale, and step your right foot forward between your hands, aligning your right knee over your heel.

b

2 Lower your left knee to the floor, and, keeping your right knee fixed in place, slide your left back until you feel a comfortable stretch in the front of your left thigh and groin. Rest the top of your left foot on the floor.

3 Inhale, and lift your torso to an upright position. At the same time, sweep your arms out to the sides and up toward the ceiling. Draw your tailbone down toward the floor, and lift your pubis toward your navel.

c

4 Tilt your head, and gaze upward while reaching your pinkies toward the ceiling. Hold for 1 minute.

5 Exhale, and fold your torso back down to your right thigh. Place your hands on the floor, and flip your toes so that the bottoms press against the floor. Exhale, and lift your left knee off the floor and step back to the Downward-Facing Dog Pose. Repeat on the other side.

DO IT RIGHT
- If your lowered knee feels uncomfortable, place a folded towel underneath it.

AVOID
- Dropping your knee to the inside or outside—it should remain forward, directly in front of you.

PRONUNCIATION & MEANING
- Anjaneyasana
- Anjaneya = a name for Hanuman, a Hindu deity who wears the crescent moon in his hair
- Also called Crescent Moon Pose, Split Leg Pose, or Kneeling Lunge

LEVEL
- Beginner

BENEFITS
- Relieves sciatica
- Tones hip abductors
- Strengthens arms and shoulders
- Stretches knee muscles, tendons, and ligaments

CONTRA-INDICATIONS & CAUTIONS
- Heart problems

BEST FOR

- rectus femoris
- obliquus internus
- obliquus externus
- biceps femoris
- deltoideus
- trapezius
- sartorius
- adductor magnus
- iliopsoas
- iliacus

ANNOTATION KEY

Black text indicates strengthening muscles

Gray text indicates stretching muscles

* indicates deep muscles

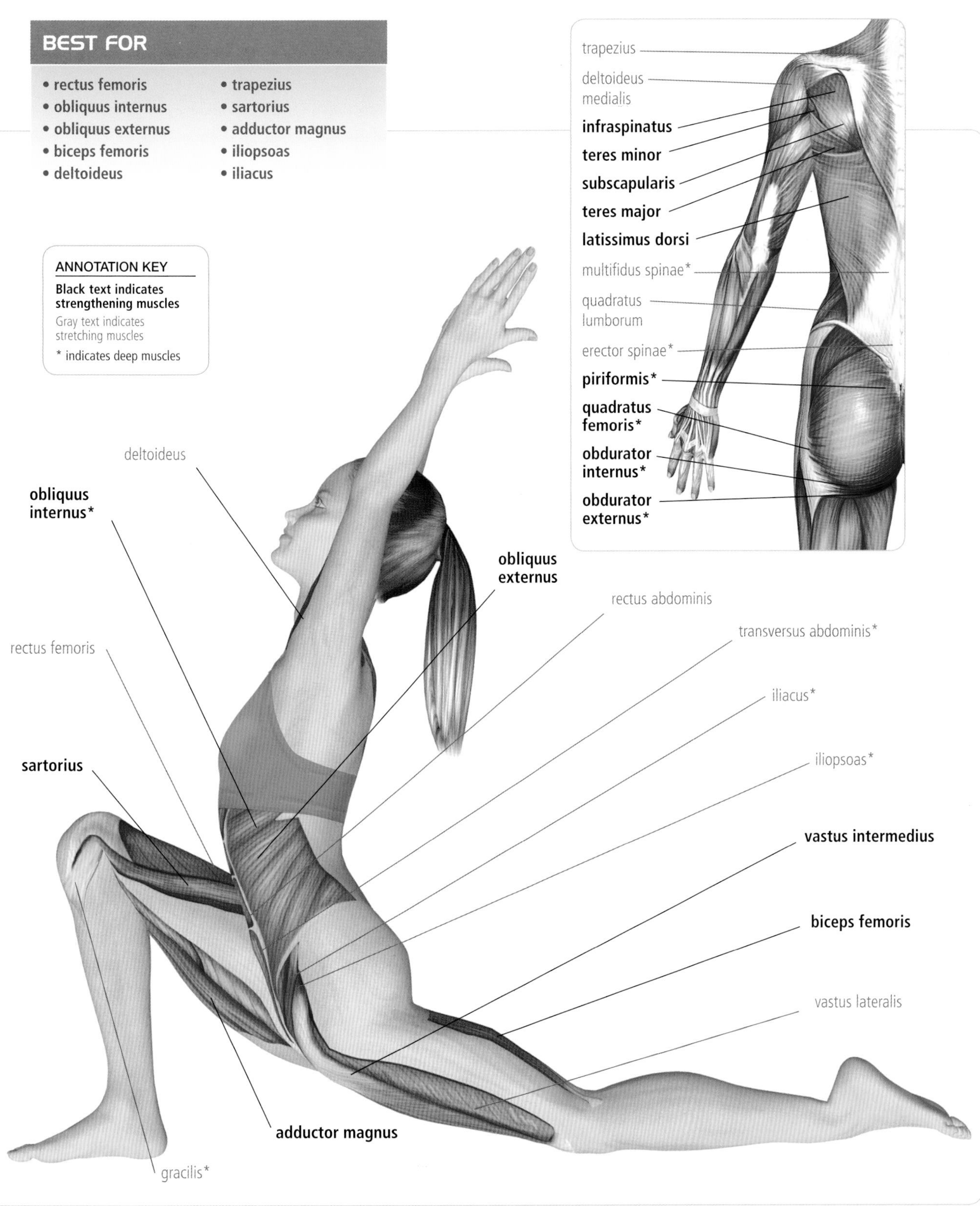

HIGH LUNGE

a

1. Stand in Mountain Pose (Tadasana, see page 32), and inhale deeply. Exhale, and carefully step back with your left leg, keeping it in line with your hips as you step back. The ball of your left foot should be in contact with the floor as you do the motion.

2. Slowly slide your left foot farther back, while bending your right knee, stacking it directly above your ankle.

3. Position your palms or fingers on the floor on either side of your right leg, and slowly press your palms or fingers against the floor to enhance the placement of your upper body and your head.

b

4. Lift your head, and gaze straight forward, while leaning your upper body forward and carefully rolling your shoulders down and backward.

5. Press the ball of your left foot gradually on the floor, contract your thigh muscles, and press up to maintain your left leg in a straight position.

6. Hold for 5 to 6 seconds. Slowly return to Mountain Pose, and then repeat on the other side.

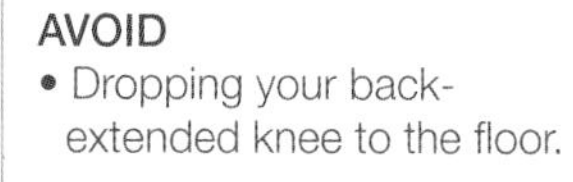

AVOID
- Dropping your back-extended knee to the floor.

PRONUNCIATION & MEANING
- There is no agreed-upon Sanskrit name for this pose.
- Sometimes called the Horse Rider's Pose (Ashva Sanchalanasana)

LEVEL
- Intermediate

BENEFITS
- Strengthens legs and arms
- Stretches groins
- Relieves constipation

CONTRA-INDICATIONS & CAUTIONS
- Arm injury
- Shoulder injury
- Hip injury
- High or low blood pressure
- Severe headache

DO IT RIGHT
- Maintain proper position of your shoulders and your whole upper body to lengthen your spine.

BEST FOR

- biceps femoris
- adductor longus
- adductor magnus
- gastrocnemius
- tibialis posterior
- iliopsoas
- biceps femoris
- rectus femoris

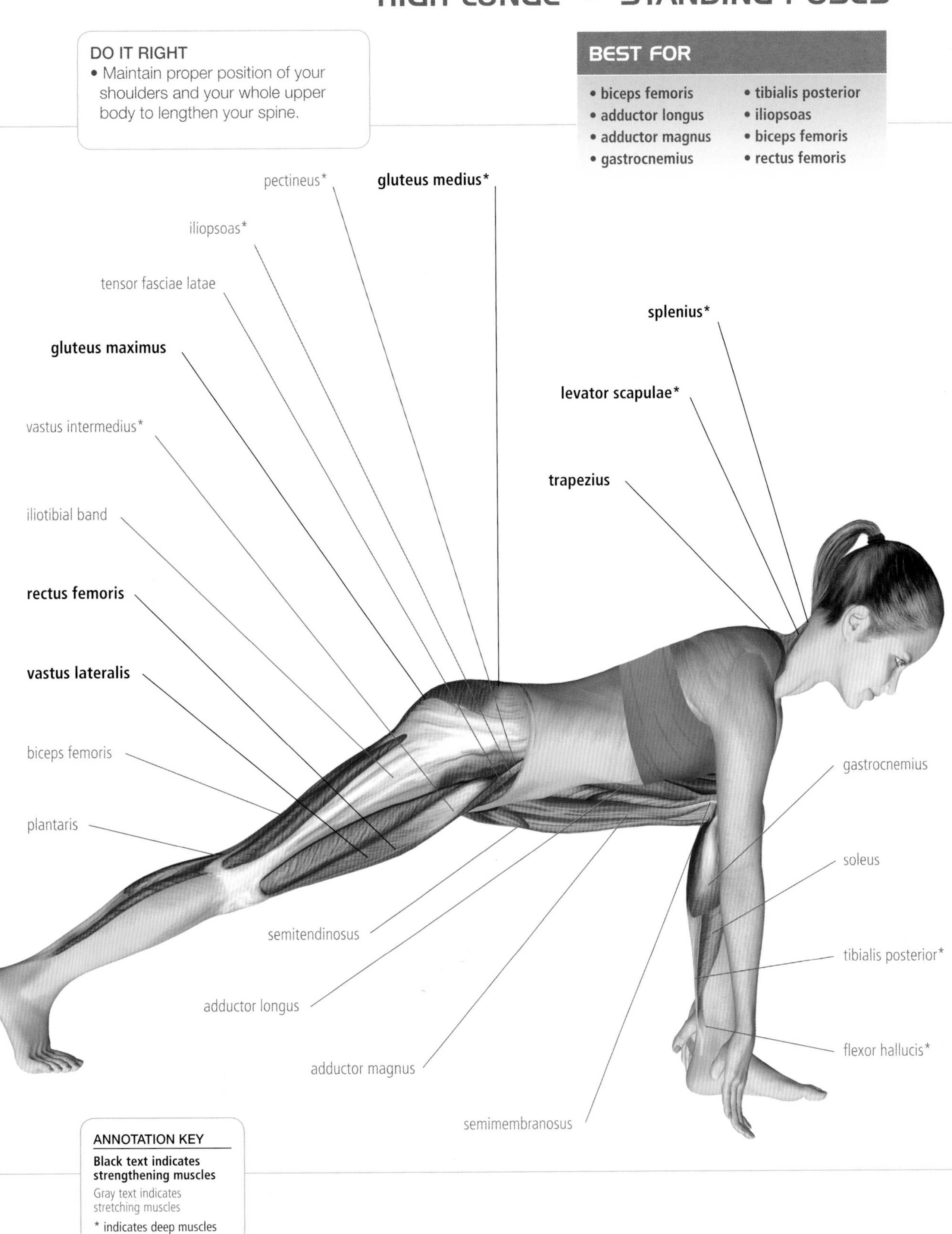

ANNOTATION KEY
Black text indicates strengthening muscles
Gray text indicates stretching muscles
* indicates deep muscles

WARRIOR POSE I

(VIRABHADRASANA I)

1 Stand in Mountain Pose (Tadasana, see page 32). Exhale, and step your left foot back 3.5 to 4 feet apart. Align your left heel behind the right heel, and then turn your left foot out 45 degrees, keeping you right foot facing straight forward. Rotate your hips so both hipbones are squared forward and are parallel to the front of your mat.

2 Inhale, and raise your arms up toward the ceiling while keeping them parallel to each other and shoulder-width apart. Firm your shoulder blades against your back, and draw them down toward your tailbone.

3 Exhale, contract yours abdominals, and tuck your tailbone under. With your left heel firmly grounded, exhale, and then slowly bend your right knee, stacking it over your heel. Your right shin should be perpendicular to the floor and your right thigh parallel to the floor.

4 Keep your head in a neutral position, gazing forward, or tilt it back and look up at your thumbs. Hold for 30 seconds to 1 minute.

5 To come up, inhale, press the back heel firmly into the floor and reach up with your arms, straightening your right knee. Turn your feet forward, exhale, and release your arms. Take a few breaths, turn your feet to the left, and repeat on the other side.

DO IT RIGHT
- Apply slightly more pressure in your right heel rather than in your toes to keep your right knee stable.
- If you are a beginner, to maintain balance, decrease the distance between your feet by several inches, still keeping your right knee over your heel.

AVOID
- Shifting your weight too far forward so that your front knee is aligned over your toes.
- Allowing your hips to shift to either side.

PRONUNCIATION & MEANING
- Virabhadrasana I (veer-ah-bah-DRAHS-anna)
- Virabhadra = the name of a fierce warrior
- Also known as Virabhadra's Pose

LEVEL
- Beginner

BENEFITS
- Strengthens arms, shoulders, thighs, ankles, and back
- Stretches hip flexors, abdominals, and ankles
- Expands chest, lungs, and shoulders
- Develops stamina
- Improves sense of balance

CONTRA-INDICATIONS & CAUTIONS
- Heart problems
- High blood pressure
- Shoulder injury

BEST FOR

- rectus abdominis
- obliquus internus
- transversus abdominis
- biceps femoris
- sartorius
- obliquus externus

ANNOTATION KEY

Black text indicates strengthening muscles

Gray text indicates stretching muscles

* indicates deep muscles

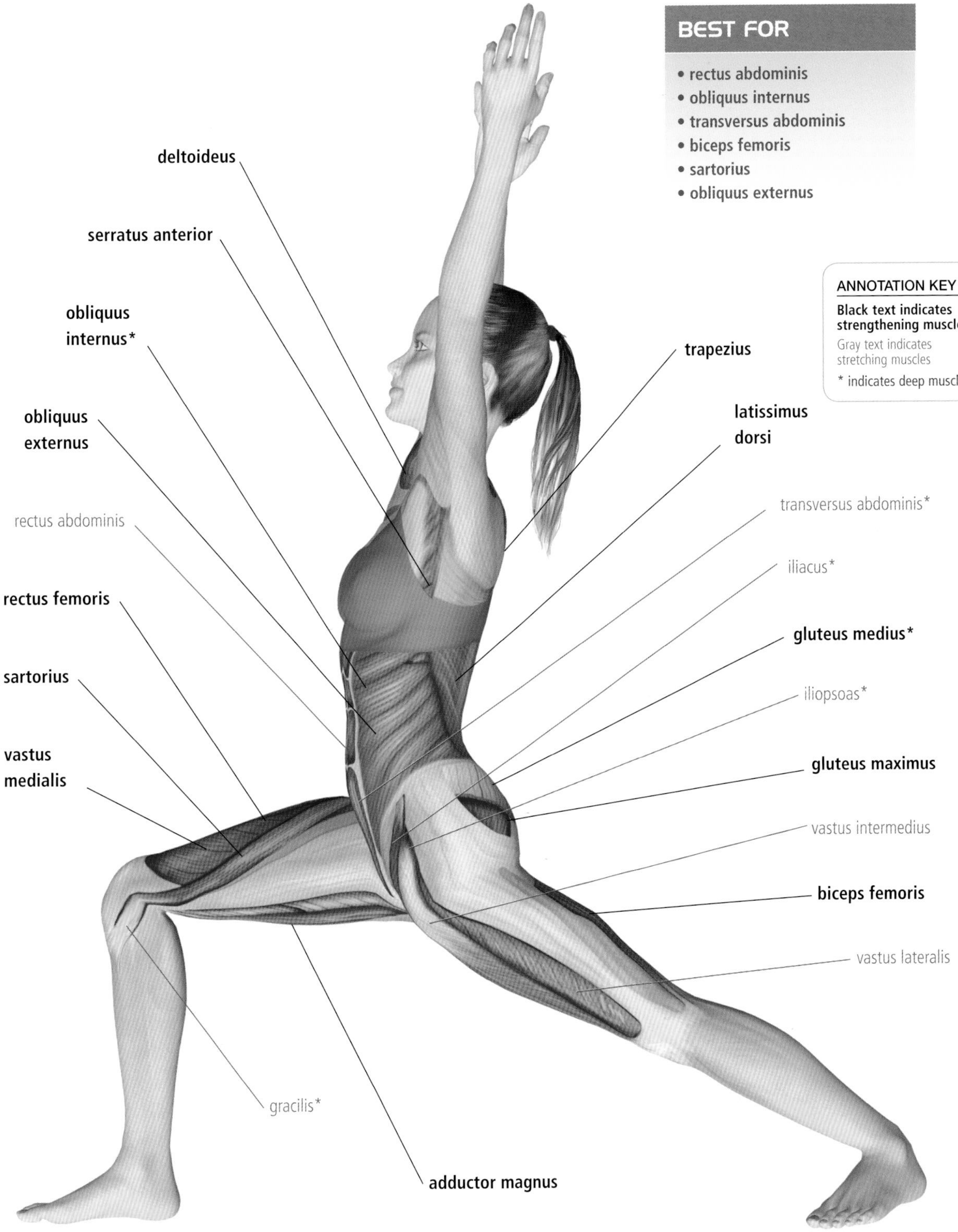

deltoideus

serratus anterior

obliquus internus*

obliquus externus

rectus abdominis

rectus femoris

sartorius

vastus medialis

gracilis*

adductor magnus

trapezius

latissimus dorsi

transversus abdominis*

iliacus*

gluteus medius*

iliopsoas*

gluteus maximus

vastus intermedius

biceps femoris

vastus lateralis

WARRIOR POSE II
(VIRABHADRASANA II)

1 Stand in Mountain Pose (Tadasana, see page 32). Exhale, and step sideways so that your feet are 3.5 to 4 feet apart.

2 Raise your arms parallel to the floor and reach them out to the sides, shoulder blades wide, palms facing downward.

a

DO IT RIGHT
- Focus on turning the knee of your bent leg outward, opening your hips and groins.

3 Turn your left foot in slightly to the right and your right foot out to the right 90 degrees. Align your right heel with your left heel. Firm your thighs and turn your right thigh outward so that the center of your right kneecap is in line with the center of your right ankle.

4 Exhale, and bend your right knee, so that your shin is perpendicular to the floor. Bring your right thigh parallel to the floor, anchoring your right knee by contracting the muscles of your left leg and pressing the outside of your left heel firmly to the floor. Keep the sides of your torso equally long and your shoulders aligned directly over your pelvis. Press your tailbone slightly toward your pubis.

5 Turn your head to the right and look out over your fingers.

6 Hold for 30 seconds to 1 minute. Inhale, and return to Mountain Pose. Reverse your feet and repeat on the other side.

b

PRONUNCIATION & MEANING
- Virabhadrasana II (veer-ah-bah-DRAHS-anna)
- Virabhadra = the name of a fierce warrior

LEVEL
- Beginner

BENEFITS
- Strengthens legs and ankles
- Stretches legs, ankles, groins, chest, and shoulders
- Stimulates digestion
- Increases stamina
- Relieves backache
- Relieves carpal tunnel syndrome
- Relieves sciatica

CONTRA-INDICATIONS & CAUTIONS
- Diarrhea
- High blood pressure
- Neck issues

AVOID
- Allowing the knee to drift over to either side.
- Leaning your torso over your bent leg.

BEST FOR

- gluteus maximus
- gluteus medius
- obliquus externus
- biceps femoris
- sartorius
- adductor longus
- adductor magnus
- sartorius

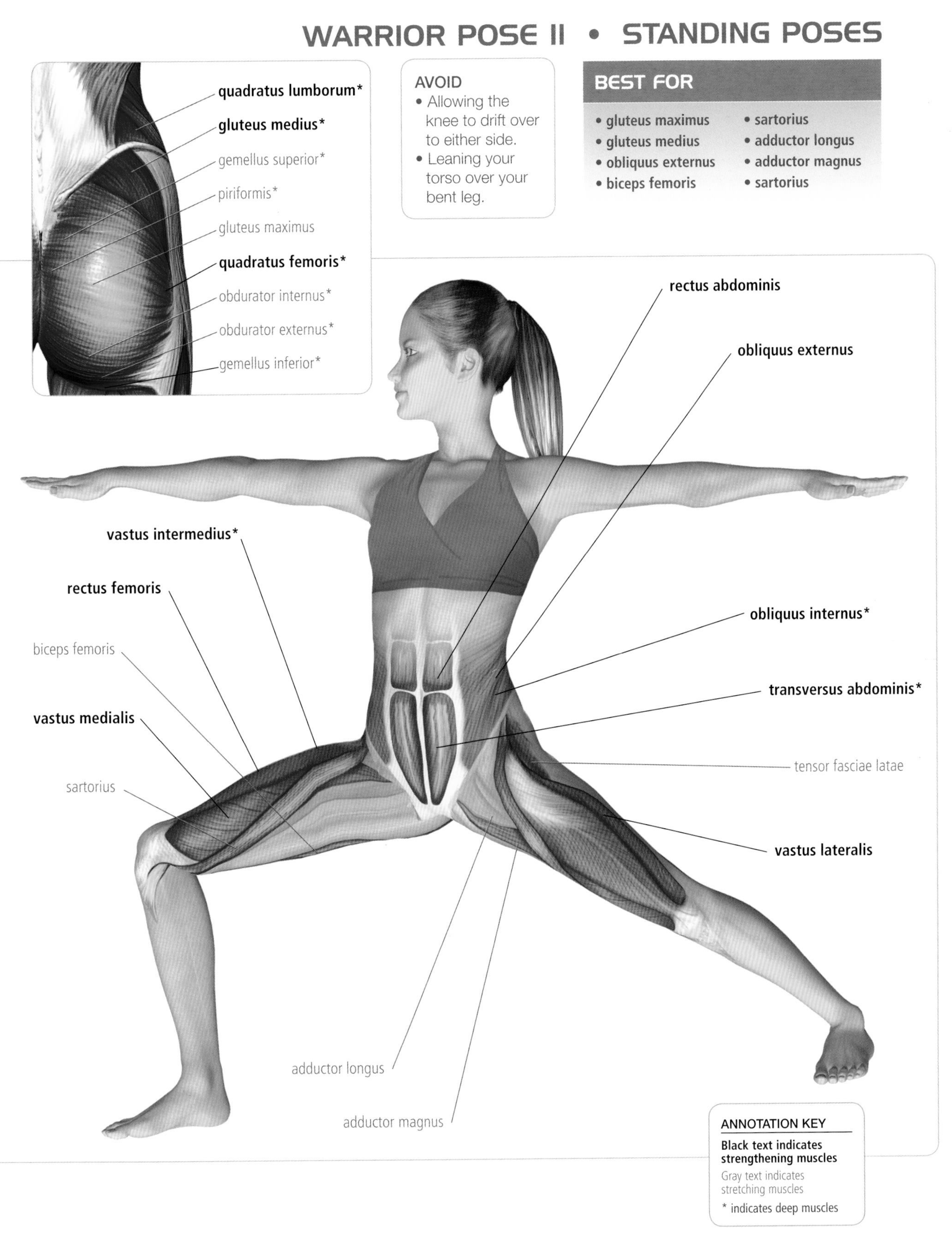

ANNOTATION KEY
Black text indicates strengthening muscles
Gray text indicates stretching muscles
* indicates deep muscles

WARRIOR POSE III

(VIRABHADRASANA III)

a

1. Stand in Mountain Pose (Tadasana, see page 32). Exhale, and step your right foot 1 foot forward, and shift all of your weight onto your right leg.

2. Inhale, and raise your arms over your head, interlacing your fingers and pointing your index fingers upward.

3. Exhale, and lift your left leg up behind you, hinging at your hips to lower your arms and torso down toward the floor.

4. Gaze down at a point on the floor for balance. Elongate your body from your left toes through the crown of your head to your fingers, making one straight line.

b

5. Hold for 30 seconds to 1 minute.

6. Inhale, and raise your arms upward as you lower your left leg back to the floor. Bring both feet together into the Mountain Pose.

7. Repeat on the other side.

DO IT RIGHT
- Position your arms, torso, and raised leg relatively parallel to the floor.

PRONUNCIATION & MEANING
- Virabhadrasana III (veer-ah-bah-DRAHS-anna)
- Virabhadra = the name of a fierce warrior

LEVEL
- Intermediate

BENEFITS
- Strengthens ankles, legs, shoulders, and back muscles
- Tones abdominals
- Improves sense of balance
- Improves posture

CONTRA-INDICATIONS & CAUTIONS
- High blood pressure

AVOID
- Tilting your pelvis so that your hips are not aligned.
- Compressing the back of your neck.

BEST FOR

- rectus abdominis
- obliquus internus
- transversus abdominis
- biceps femoris
- erector spinae
- gluteus maximus
- deltoideus posterior

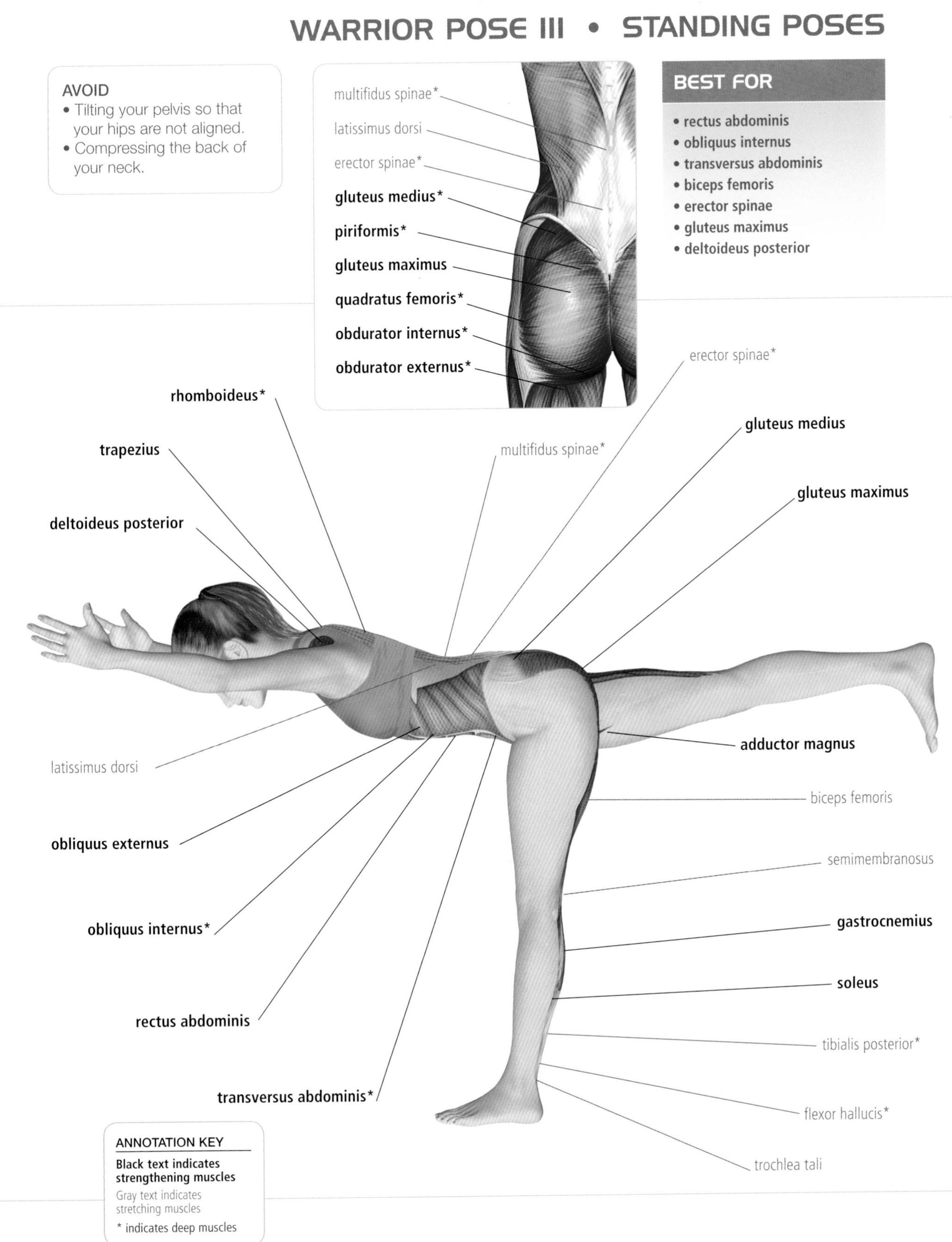

ANNOTATION KEY
Black text indicates strengthening muscles
Gray text indicates stretching muscles
* indicates deep muscles

EXTENDED SIDE ANGLE POSE

(UTTHITA PARSVAKONASANA)

a

1 Stand in Warrior II Pose (Virabhadrasana II, see pages 56–57), with your right leg bent, your left leg extended, and your arms raised to the sides, parallel to the floor.

2 Anchor your left heel to the floor. Your right knee should be bent over your right ankle, so that your shin is perpendicular to the floor. Aim the inside of your knee toward the outside of your foot. Bring your right thigh parallel to the floor.

AVOID
- Sagging at the middle—your forward thigh should remain parallel to the floor.
- Lifting the heel of your extended leg.

b

3 Firm your shoulder blades against your back ribs. Extend your left arm straight up toward the ceiling, and then turn your left palm to face toward your head. Inhale, and reach your left arm over the back of your left ear, palm facing the floor, stretching from your left heel through your left fingertips to lengthen the entire left side of your body. Make sure your elbow remains straight.

4 Turn your head to gaze at your left arm. Release your right shoulder away from the ear, creating as much length along the right side of your torso as you do along the left.

5 Continue to ground your left heel to the floor, exhale, and lay the right side of your torso down onto the top of your right thigh. Press your right fingertips or palm on the floor just outside of your right foot. Push your right knee back against your inner arm, while tucking your tailbone toward your pubis and pressing your hips forward.

6 Hold for 30 seconds to 1 minute.

7 Inhale, and begin to rise. Push both heels strongly into the floor, and reach your left arm toward the ceiling to lighten the upward movement. Reverse your feet, and repeat on the other side.

PRONUNCIATION & MEANING
- Utthita Parsvakonasana (oo-TEE-tah parsh-vah-cone-AHS-anna)
- *utthita* = extended; *parsva* = side, flank; *kona* = angle

LEVEL
- Beginner

BENEFITS
- Strengthens legs, knees, and ankles
- Stretches legs, knees, ankles, groins, spine, waist, chest and lungs, and shoulders
- Stimulates abdominal organs
- Increases stamina

CONTRA-INDICATIONS & CAUTIONS
- Headache
- Insomnia
- High or low blood pressure

DO IT RIGHT

- If you feel unsteady, brace your back heel against a wall.
- If you have trouble reaching the floor with your hand, place your right hand on a block, or bend your elbow and place your forearm on your right thigh, hand facing up, and shoulder still away from ear.

BEST FOR

- semitendinosus
- semimembranosus
- obliquus internus
- transversus abdominis
- biceps femoris
- sartorius
- obliquus externus
- piriformis
- gracilis
- tensor fasciae latae

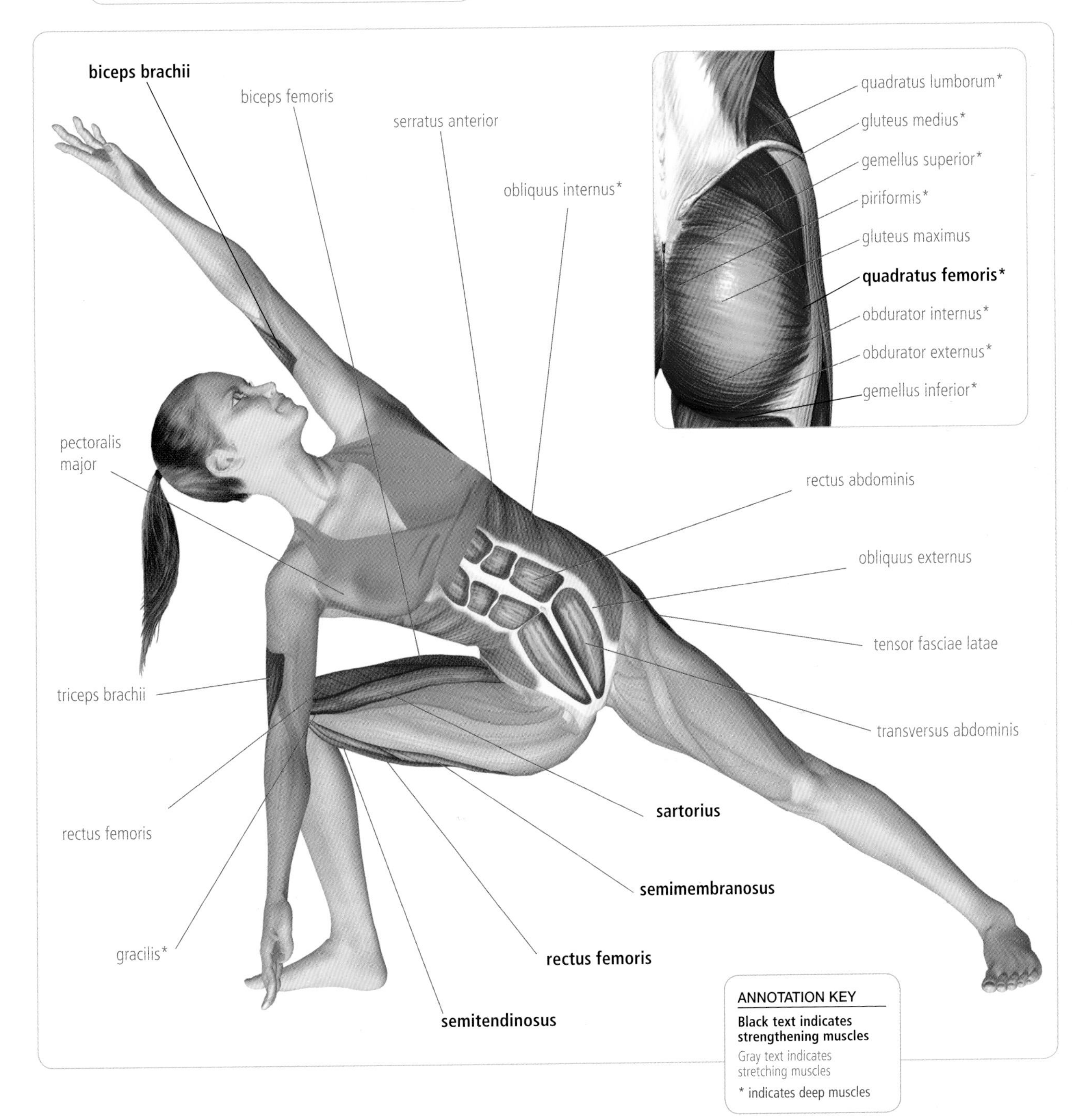

ANNOTATION KEY
Black text indicates strengthening muscles
Gray text indicates stretching muscles
* indicates deep muscles

FORWARD BENDS

Forward bends may be as straightforward as yoga poses come, but they are certainly not lacking in variety. Both seated and standing, forward bends include poses with the legs together and separated—either out to the side or in opposition to each other.

All of the poses in this section will challenge your body's sense of alignment. Forward bends will stretch your hamstrings and the entire back of your body, releasing your spine. It is important to bend at your hips rather than your waist, because bending at your waist will shorten your movement and strain your back. To do this, flatten your back and fold—don't curl—into the pose.

INTENSE SIDE STRETCH

(PARSVOTTANASANA)

a

1 Stand in Mountain Pose (Tadasana, see page 32). To assume Reverse Prayer, or Paschima Namaskar, bring your hands together behind your back. Bend your knees slightly, and round your torso forward. Touch your fingertips together, and rotate them inward toward the center of your shoulder blades. Once your fingers are pointing upward with your hands parallel to your spine, stand up tall, and draw your elbows forward, keeping your shoulders down.

AVOID

- Lifting your back heel off the floor.
- Rounding your spine to lower your torso to your front leg.
- Turning your hips out to either side.

b

2 Exhale, and take one large step forward with your right leg, 3 to 4 feet away from your left. Turn your back foot out slightly, and keep your right foot pointed forward. Square your hips forward by rotating your torso slightly to the right, and tuck your tailbone toward your pubis. Press your left heel to the floor, and contract your leg muscles. Lift up tall through your spine and chest.

DO IT RIGHT

- If your shoulders aren't flexible enough to do this pose in Reverse Prayer, place your hands on the floor or bend your arms behind your back and cross them, holding each elbow in your opposite hand.

c

3 Begin to lean forward with your torso with another exhalation, keeping your back flat. Stretch forward until your torso is parallel to the floor. Make sure your leg muscles are still contracted and your feet firmly grounded.

PRONUNCIATION & MEANING

- Parsvottanasana (parsh-voh-tahn-AHS-anna)
- *parsva* = side, flank; *ut* = intense; *tan* = to stretch or extend
- Also called Revolved Side Stretch

LEVEL

- Intermediate

BENEFITS

- Stretches shoulders, spine, and hamstrings
- Strengthens legs
- Stimulates digestion

CONTRA-INDICATIONS & CAUTIONS

- High blood pressure
- Back injury

4 With your back flat, draw your torso down to your right thigh.

5 Hold for 15 to 30 seconds. Repeat with your left leg in front.

d

BEST FOR

- biceps femoris
- semitendinosus
- gluteus medius
- gluteus maximus
- gastrocnemius
- soleus
- deltoideus

trapezius
deltoideus medialis
infraspinatus
teres minor
subscapularis
teres major
latissimus dorsi
quadratus lumborum
erector spinae*

gluteus medius*
erector spinae*
iliopsoas*
gluteus maximus
biceps femoris
semitendinosus
vastus medialis
sartorius
gastrocnemius
soleus
tibialis posterior*
rectus femoris
vastus lateralis
latissimus dorsi
deltoideus

ANNOTATION KEY
Black text indicates strengthening muscles
Gray text indicates stretching muscles
* indicates deep muscles

STANDING FORWARD BEND TO

(UTTANASANA)

a

1. Stand in Mountain Pose (Tadasana, see page 32), and then raise your arms toward the ceiling in Upward Salute (Urdhva Hastasana, see page 36).

2. Exhale, and bend forward from your hips, sweeping your arms to the sides with your palms facing the floor. While you lower your torso, keep your back flat, and tuck your abdominals in toward your spine. Lengthen your spine as much as possible.

AVOID
- Rolling your spine into or out of the pose.
- Compressing the back of your neck as you look forward.

3. Fold your torso and abdominals onto the front of your legs, aiming your forehead toward your shins. Grasp the backs of your ankles, and contract your thigh muscles to try to straighten your knees as much as possible.

4. With each exhalation, draw your sit bones up to the ceiling, and elongate your spine to the floor even more to create a deeper stretch.

5. Hold for 30 seconds to 1 minute.

b

DO IT RIGHT
- If you have tight hamstrings, bend your knees as you fold your torso forward. Work on pressing your knees straight once you are in the forward bend. You may also bend your knees on your way up to Standing Half Forward Bend to help you create a slight arch in your back.

PRONUNCIATION & MEANING
- Uttanasana (oot-tan-AHS-anna)
- *ut* = intense; *tan* = to stretch or extend
- Ardha Uttanasana (are-dah oot-tan-AHS-anna)
- *ardha* = half; *ut* = intense; *tan* = to stretch or extend

LEVEL
- Beginner

BENEFITS
- Stretches spine, hamstrings, calves, and hips
- Strengthens spine and thighs
- Improves posture
- Relieves stress

CONTRA-INDICATIONS & CAUTIONS
- Back injury
- Neck injury
- Osteoporosis

STANDING HALF FORWARD BEND
(ARDHA UTTANASANA)

BEST FOR

- biceps femoris
- iliotibial band
- gluteus maximus
- gluteus medius
- erector spinae

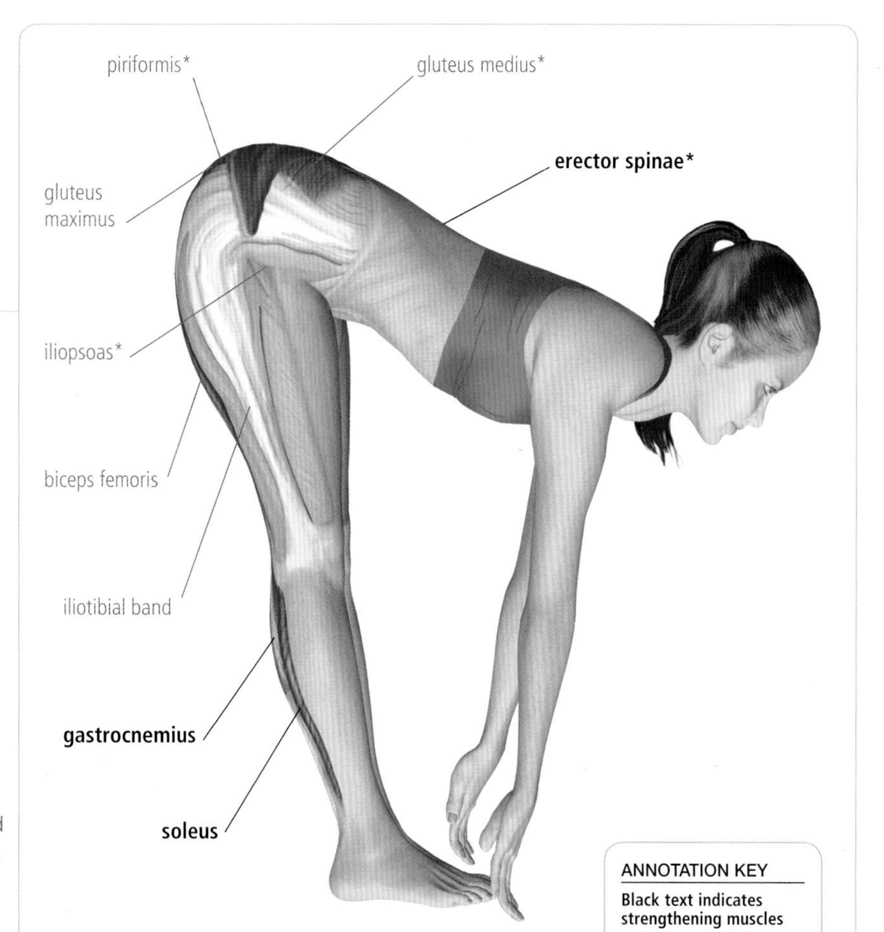

ANNOTATION KEY

Black text indicates strengthening muscles

Gray text indicates stretching muscles

* indicates deep muscles

6 From Standing Forward Bend, move into Standing Half Forward Bend (Ardha Uttanasana) by placing your hands beside your feet. Inhale, and lift your head and upper torso away from your legs. Your back should be flat. Straighten your elbows and use your fingertips to guide your lift.

7 Lift your chest forward, and elongate your spine into a slight arch. Lengthen the back of your neck as you gaze forward.

8 Hold for 10 to 30 seconds. Lower yourself back down to Standing Forward Bend, or inhale, and lift your torso all the way back up to Mountain Pose.

HEAD-TO-KNEE FORWARD BEND
(JANU SIRSASANA)

a

❶ Begin in Staff Pose (Dandasana, see page 23). Bend your left knee, and draw your heel toward your groin, placing the sole of your foot on your right inner thigh. Lower your left knee to the floor. Your right leg should sit at a right angle to your left shin. Draw both sit bones to the floor.

❷ Inhale, and lift up through your spine. Turn your torso slightly to your right as you exhale so that it aligns with your right leg. Flex your foot, and contract the muscles in your right thigh to push the back of your leg toward the floor.

❸ With another exhalation, stretch your sternum forward as you fold your torso over your right leg. Grasp the inside of your right foot with your left hand. Use your right hand to guide your torso to the right.

❹ Extend your right arm forward toward your right foot. You may grasp your foot with both hands or place your hands on the floor on either side of your foot with your elbows bent. If possible, place your forehead on your right shin. With each inhalation, lengthen your spine, and with each exhalation, deepen the stretch.

❺ Hold for 1 to 3 minutes. Repeat with your left leg straight and your right leg bent.

b

BEST FOR
- **biceps femoris**
- **gastrocnemius**
- **semimembranosus**
- **quadratus femoris**
- **iliotibial band**
- **latissimus dorsi**

PRONUNCIATION & MEANING
- Janu Sirsasana (JAH-new shear-SHAHS-anna)
- *janu* = knee; *sirsa* = head

LEVEL
- Beginner

BENEFITS
- Stretches hamstrings, groins, and spine
- Stimulates digestion
- Relieves headaches
- Alleviates high blood pressure

CONTRA-INDICATIONS & CAUTIONS
- Knee injury
- Lower-back injury
- Diarrhea

AVOID
- Allowing the foot of your bent leg to shift beneath your straight leg.

DO IT RIGHT
- While bending forward, your abdominals should be the first parts of your body to touch your thigh. Your head should be the last.

ANNOTATION KEY
Black text indicates strengthening muscles

Gray text indicates stretching muscles

* indicates deep muscles

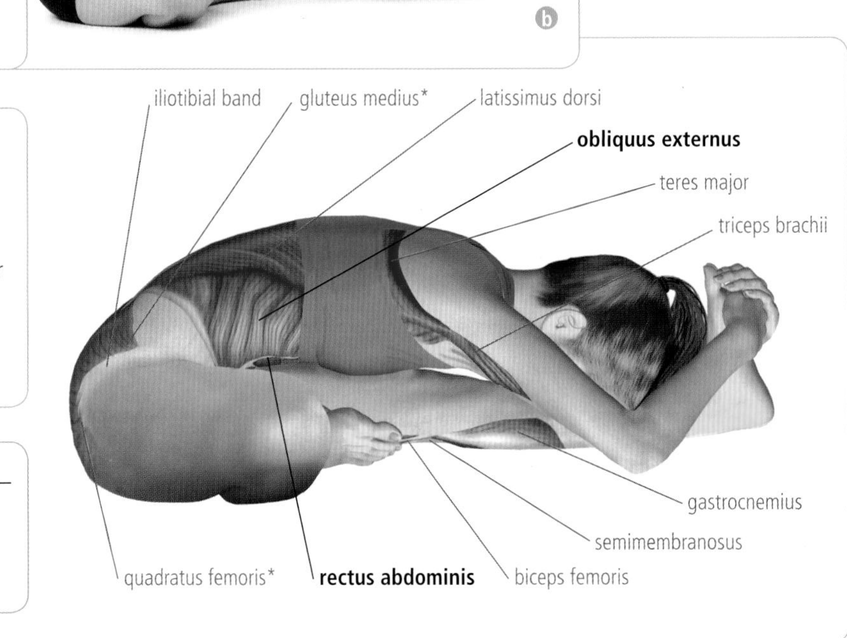

SEATED FORWARD BEND

(PASCHIMOTTANASANA)

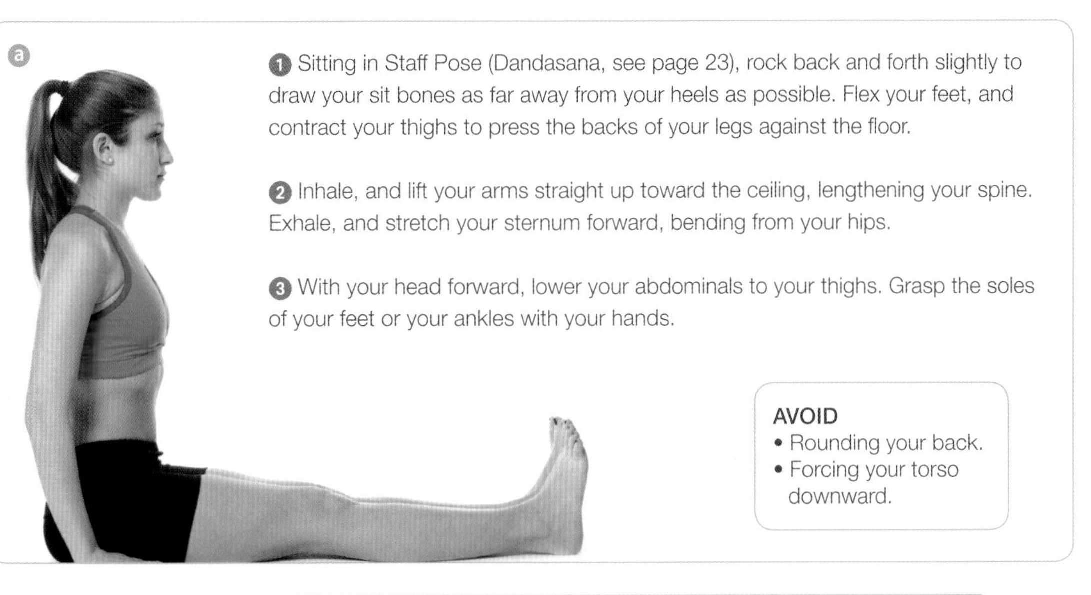

1. Sitting in Staff Pose (Dandasana, see page 23), rock back and forth slightly to draw your sit bones as far away from your heels as possible. Flex your feet, and contract your thighs to press the backs of your legs against the floor.

2. Inhale, and lift your arms straight up toward the ceiling, lengthening your spine. Exhale, and stretch your sternum forward, bending from your hips.

3. With your head forward, lower your abdominals to your thighs. Grasp the soles of your feet or your ankles with your hands.

AVOID
- Rounding your back.
- Forcing your torso downward.

BEST FOR
- biceps femoris
- semitendinosus
- semimembranosus
- quadratus femoris
- erector spinae
- obdurator externus

4. With each inhalation, lengthen your spine. With each exhalation, deepen the stretch. If possible, bend your elbows to gently lengthen your torso forward, and place your forehead on your shins.

5. Hold for 1 to 3 minutes.

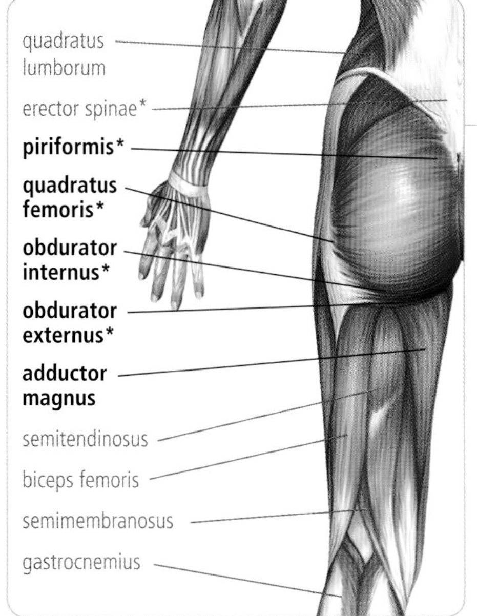

DO IT RIGHT
- To help guide the forward bend from your hips, place a folded blanket beneath your buttocks.
- Lengthen your spine from your hips to your neck.

PRONUNCIATION & MEANING
- Paschimottanasana (POSH-ee-moh-tan-AHS-anna)
- *pascha* = behind, west, after; *uttana* = intense stretch

LEVEL
- Beginner

BENEFITS
- Stretches hamstrings, shoulders, and spine
- Stimulates digestion
- Relieves headache and stress
- Alleviates high blood pressure

CONTRA-INDICATIONS & CAUTIONS
- Back injury
- Diarrhea

WIDE-LEGGED FORWARD BEND

(PRASARITA PADOTTANASANA)

1 Stand in Mountain Pose (Tadasana, see page 32). Take a large step—about 3 to 4 feet—to the side. Your feet should be parallel to each other. Lift up through your spine, and contract your thigh muscles.

2 Exhale, and bend forward from your hips, keeping your back flat. Draw your sternum forward as you lower your torso, gazing straight ahead. With your elbows straight, place your fingertips on the floor.

3 With another exhalation, place your hands on the floor in between your feet, and lower your torso into a full forward bend. Lengthen your spine by pulling your sit bones up toward the ceiling and drawing your head to the floor. If possible, bend your elbows and place your forehead on the floor.

a

AVOID

- Bending forward from your waist.
- Compressing the back of your neck as you look forward.

4 Hold for 30 seconds to 1 minute. To come out of the pose, straighten your elbows and raise your torso while keeping your back flat.

b

DO IT RIGHT

- Contract your leg muscles, and ground your feet throughout the pose.
- If you have trouble reaching your hands to the floor, widen your stance or place blocks on the floor for support.

PRONUNCIATION & MEANING

- Prasarita Padottanasana (pra-sa-REE-tah pah-doh-tahn-AHS-anna)
- *prasarita* = spread, expanded; *pada* = foot; *ut* = intense; *tan* = to stretch or extend

LEVEL

- Beginner

BENEFITS

- Stretches and strengthens hamstrings, groins, and spine

CONTRA-INDICATIONS & CAUTIONS

- Lower-back issues

WIDE-LEGGED FORWARD BEND • FORWARD BENDS

MODIFICATION

Easier: Follow step 1, and then exhale, bending forward until your torso is nearly parallel to the ground. Place your hands on the ground in line with your shoulders, making sure that your lower back is straight. Hold for 30 seconds to 1 minute.

BEST FOR

- gluteus maximus
- biceps femoris
- semitendinosus
- adductor longus
- adductor magnus
- tibialis anterior
- erector spinae

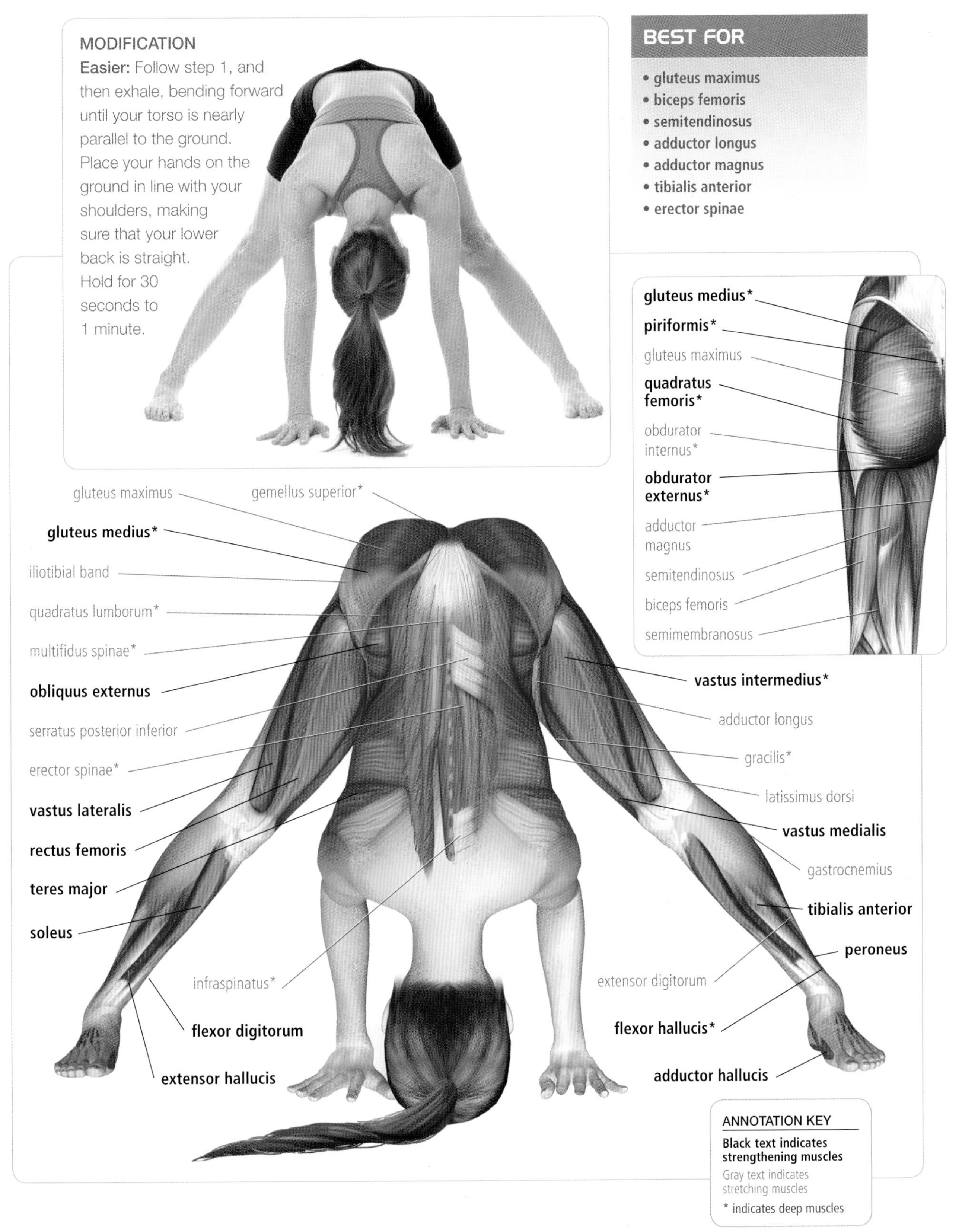

WIDE-ANGLE SEATED BEND

(UPAVISTHA KONASANA)

a

1. Sit in Staff Pose (Dandasana, see page 23).

AVOID

- Bending forward from your waist.
- Forcing your torso to the ground.

2. Separate your legs wide. Turn your thighs slightly outward so that your knees point up toward the ceiling. Flex your feet. Place your hands on the floor behind your buttocks to push them forward, separating your legs even farther.

3. Inhale, and lift up with your torso, placing your hands on the floor in front of you. Contract your leg muscles, and press the backs of your thighs and both sit bones into the floor.

b

PRONUNCIATION & MEANING

- Upavistha Konasana (oo-pah-VEESH-tah cone-AHS-anna)
- *upavistha* = seated; *kona* = angle

LEVEL

- Intermediate

BENEFITS

- Stretches groins and hamstrings
- Strengthens spine

CONTRA-INDICATIONS & CAUTIONS

- Lower-back injury

4. Exhale, and bend forward from your waist, keeping your back flat. Walk your hands in front of you to lower your torso slowly toward the floor. Gaze forward. Stretch as far as possible without rounding your back.

5. Hold for 1 to 2 minutes.

c

DO IT RIGHT
- If you have trouble bending forward from your hips or sitting with your legs open wide, place a folded blanket beneath your buttocks.
- Keep your knees pointed up toward the ceiling.

BEST FOR
- erector spinae
- piriformis
- gluteus medius
- gracilis
- semitendinosus
- semimembranosus
- biceps femoris
- adductor longus
- adductor magnus

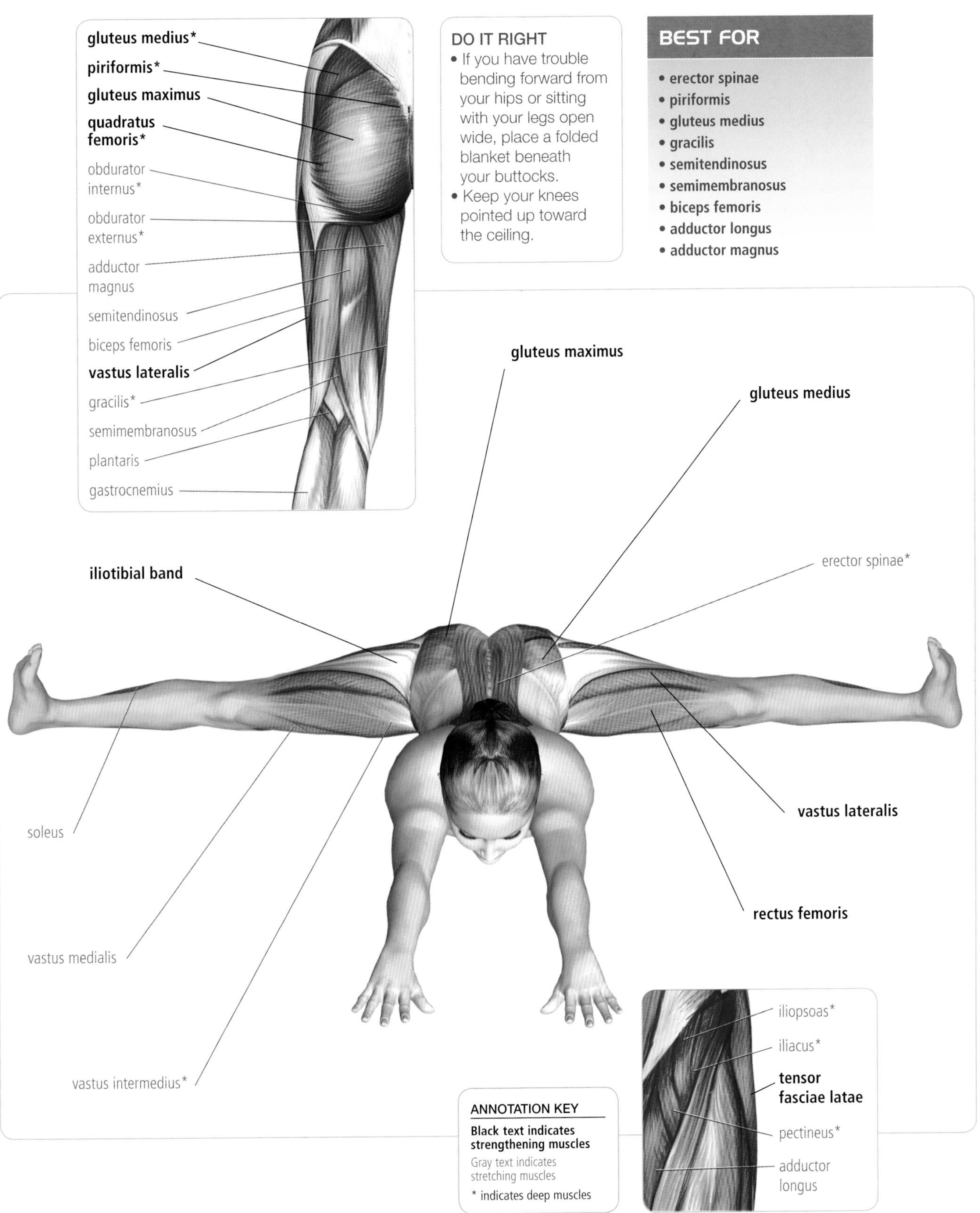

ANNOTATION KEY
Black text indicates strengthening muscles
Gray text indicates stretching muscles
* indicates deep muscles

STANDING SPLIT POSE

(URDHVA PRASARITA EKA PADASANA)

a

1 Stand in Mountain Pose (Tadasana, see page 32), and shift your weight onto your left foot.

2 Bend forward with your back flat, simultaneously raising your right leg behind you. Square your shoulders and your hips forward. Reach your fingertips toward the floor.

3 Exhale, and contract your leg muscles as you fold your torso onto your left thigh. Lift your right heel toward the ceiling, extending both legs in opposite directions.

4 Relax your shoulders down toward the floor. In this position, your left knee is pointed forward and your right knee is pointed straight behind you. If possible, grasp the back of your left ankle with your right hand. Maintain balance with your left palm on the floor.

5 Hold for 30 seconds to 1 minute. Repeat on the other side.

b

AVOID
- Rotating your standing knee inward.
- Rounding your spine.
- Bending forward from your waist.

DO IT RIGHT
- Lower your torso, and lift your back leg simultaneously.
- Tuck your chin, and elongate the back of your neck.

PRONUNCIATION & MEANING
- Urdhva Prasarita Eka Padasana (oo-pah-VEESH-tah cone-AHS-anna)
- *urdhva* = upward; *prasarita* = spread out; *eka* = one; *pada* = foot

LEVEL
- Advanced

BENEFITS
- Stretches groins, thighs, and calves
- Strengthens thighs, knees, and ankles
- Improves balance

CONTRA-INDICATIONS & CAUTIONS
- Lower-back injury
- Ankle injury
- Knee injury

BEST FOR

- biceps femoris
- semitendinosus
- sartorius
- rectus femoris
- tensor fasciae latae
- gluteus maximus
- gastrocnemius

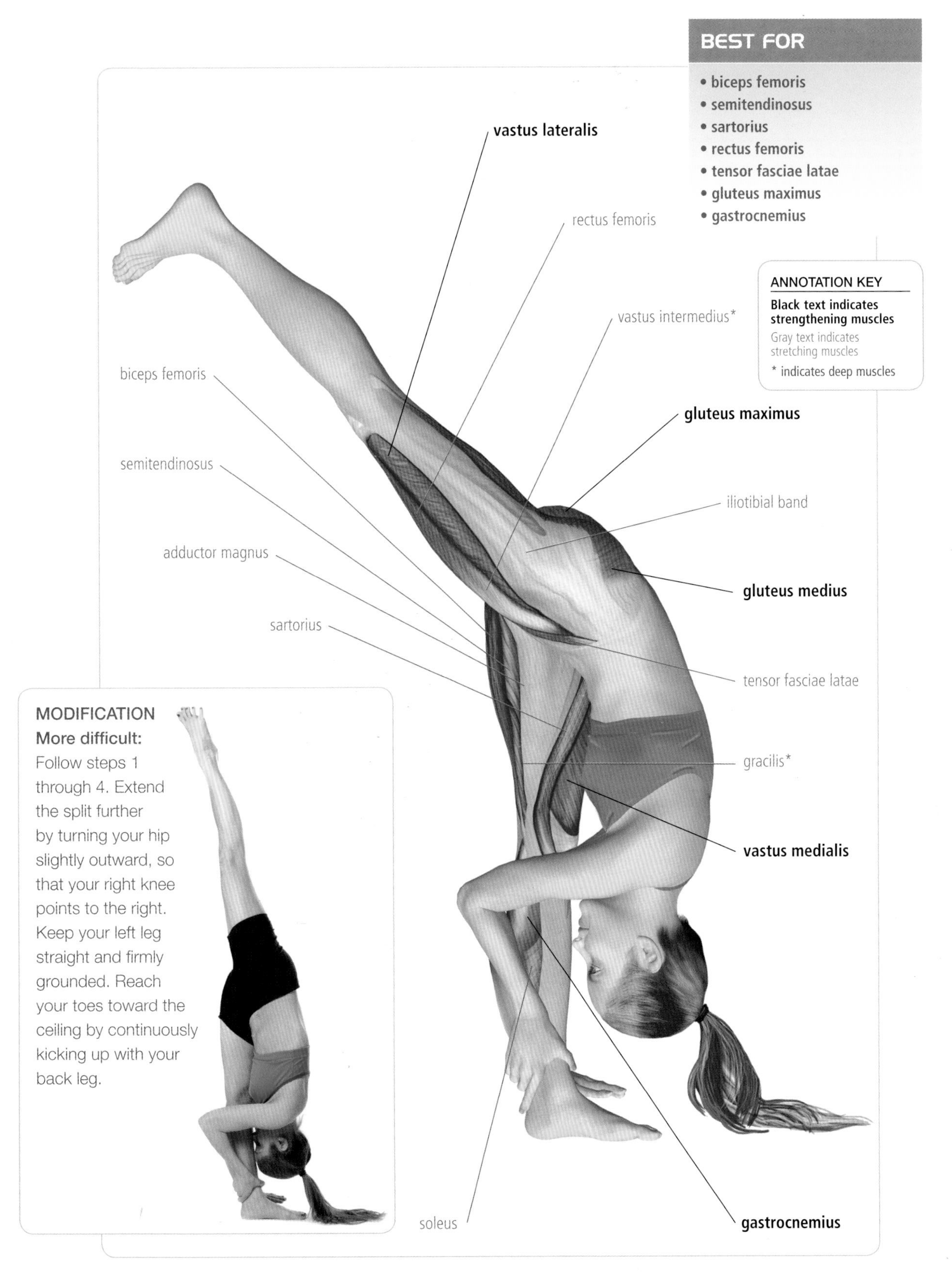

ANNOTATION KEY

Black text indicates strengthening muscles

Gray text indicates stretching muscles

* indicates deep muscles

MODIFICATION

More difficult: Follow steps 1 through 4. Extend the split further by turning your hip slightly outward, so that your right knee points to the right. Keep your left leg straight and firmly grounded. Reach your toes toward the ceiling by continuously kicking up with your back leg.

BACKBENDS

Yoga novices often view back-bending poses as unfamiliar and uncomfortable—and reasonably so. Many of us spend much of our lives bending forward or hunched over in chairs. The benefits of the back-bending poses, however, go far beyond simply improving posture. Backbends are more like full-body bends. They will stretch your shoulders, abdominals, and tops of your legs, and they will open your chest, strengthen your back, and create mobility in your hips and spine. They are invigorating and build a healthy nervous system.

It is important when entering back-bending poses that you exercise a great deal of patience. Go slowly and carefully, and do not force your body into a deeper or a more advanced pose than your muscles are ready for. Make sure that you have warmed up properly, and take special precaution if you have a chronic or recent back injury.

UPWARD-FACING DOG POSE
(URDHVA MUKHA SVANASANA)

ⓐ

❶ Lie prone on the floor. Bend your elbows, placing your hands flat on the floor on either side of your chest. Keep your elbows pulled in toward your body. Separate your legs one hip-width apart, and extend through your toes. The tops of your feet should be touching the floor.

❷ Inhale, and press against the floor with your hands and the tops of your feet, lifting your torso and hips off the floor. Contract your thighs, and tuck your tailbone toward your pubis.

ⓑ

❸ Lift through the top of your chest, fully extending your arms and creating an arch in your back from your upper torso. Push your shoulders down and back, and elongate your neck as you gaze slightly upward.

❹ Hold for 15 to 30 seconds, and exhale as you lower yourself to the floor.

DO IT RIGHT
- Elongate your legs and arms to create full extension.
- Make sure that your wrists are positioned directly below your shoulders so that you don't exert too much pressure on your lower back.

PRONUNCIATION & MEANING
- Urdhva Mukha Svanasana (OORD-vah MOO-kah shvon-AHS-anna
- *urdhva* = rising upward, elevated; *mukha* = face; *shvana* = dog

LEVEL
- Beginner

BENEFITS
- Strengthens spine, arms, and wrists
- Stretches chest and abdominals
- Improves posture

CONTRA-INDICATIONS & CAUTIONS
- Back injury
- Wrist injury or carpal tunnel syndrome

AVOID
- Lifting your shoulders up toward your ears.
- Hyperextending your elbows.
- Jutting your rib cage out of your chest.
- Dropping your thighs to the floor.

BEST FOR

- rhomboideus
- teres major
- teres minor
- trapezius
- latissimus dorsi
- erector spinae
- quadratus lumborum
- gluteus maximus
- pectoralis major
- serratus anterior
- rectus abdominis
- triceps brachii

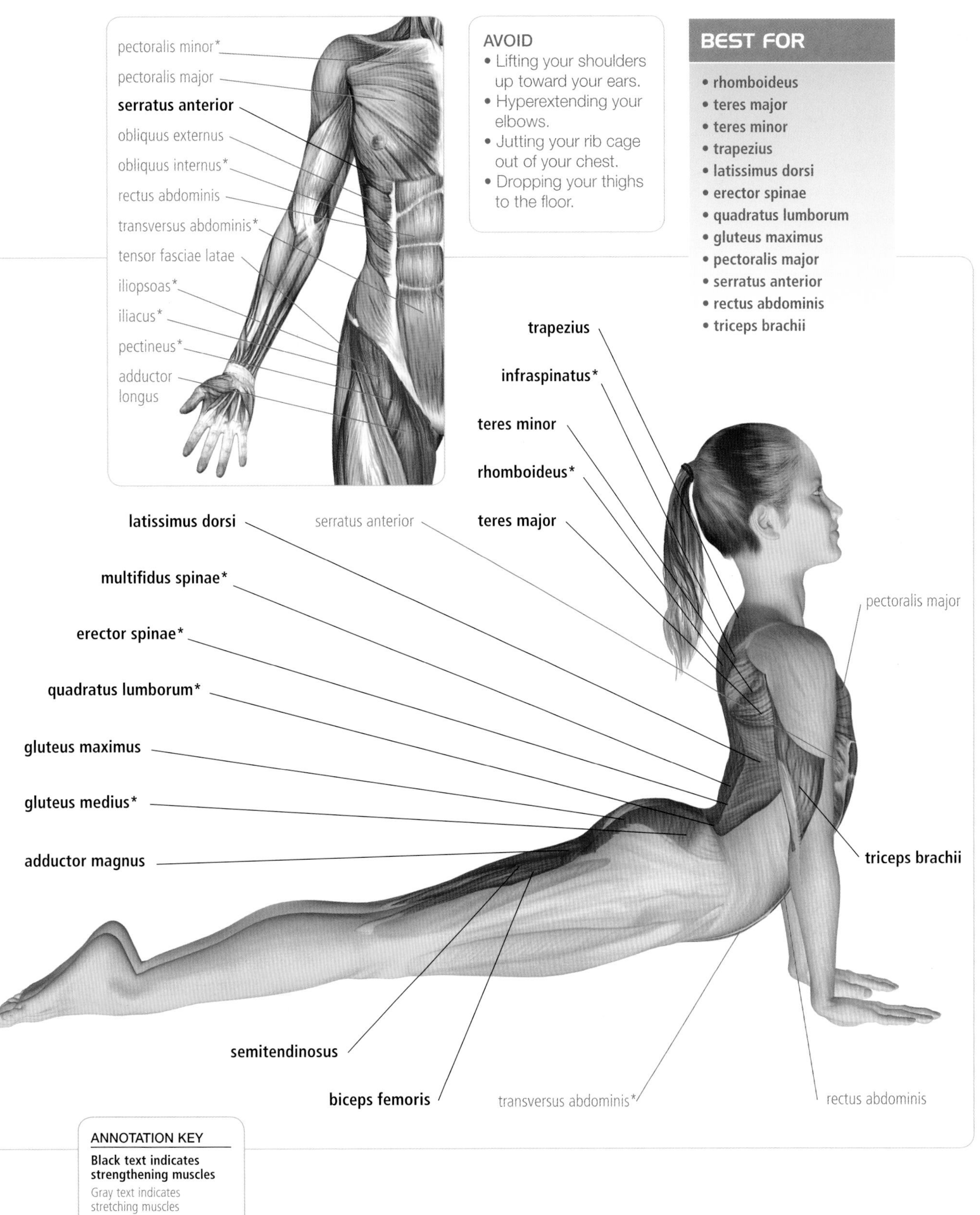

ANNOTATION KEY
Black text indicates strengthening muscles
Gray text indicates stretching muscles
* indicates deep muscles

COBRA POSE
(BHUJANGASANA)

a

1 Lie prone on the floor. Bend your elbows, placing your hands flat on the floor beside your chest. Keep your elbows pulled in toward your body. Extend your legs, pressing your pubis, thighs, and tops of your feet into the floor.

2 Inhale, and lift your chest off the floor, pushing down with your hands to guide your lift. Keep your pubis pressed against the floor.

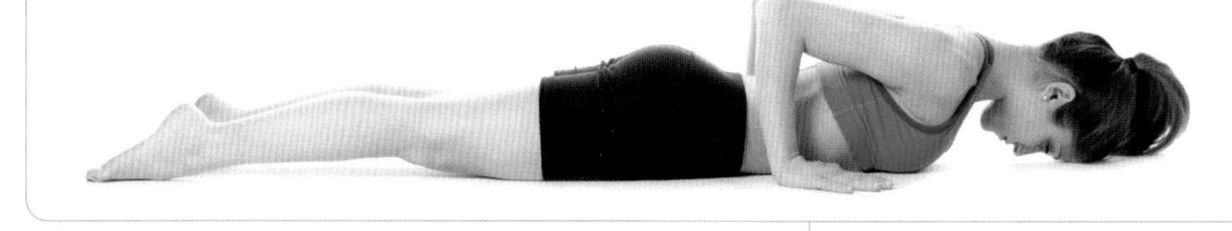

b

3 Lift through the top of your chest. Pull your tailbone down toward your pubis. Push your shoulders down and back, and elongate your neck as you gaze slightly upward.

4 Hold for 15 to 30 seconds, and exhale as you lower yourself to the floor.

PRONUNCIATION & MEANING
- Bhujangasana (boo-jang-GAHS-anna)
- *bhujang* = snake, serpent; *bhuja* = arm or shoulder; *anga* = limb

LEVEL
- Beginner

BENEFITS
- Strengthens spine and buttocks
- Stretches chest, abdominals, and shoulders

CONTRA-INDICATIONS & CAUTIONS
- Back injury

AVOID
- Tensing your buttocks, which adds pressure on your lower back.
- Splaying your elbows out to the sides.
- Lifting your hips off the floor.

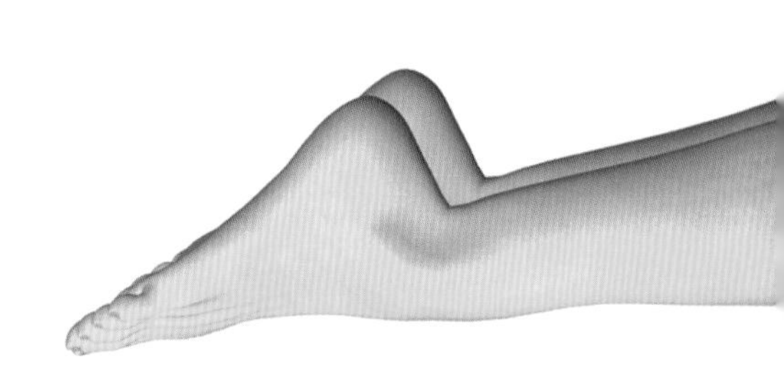

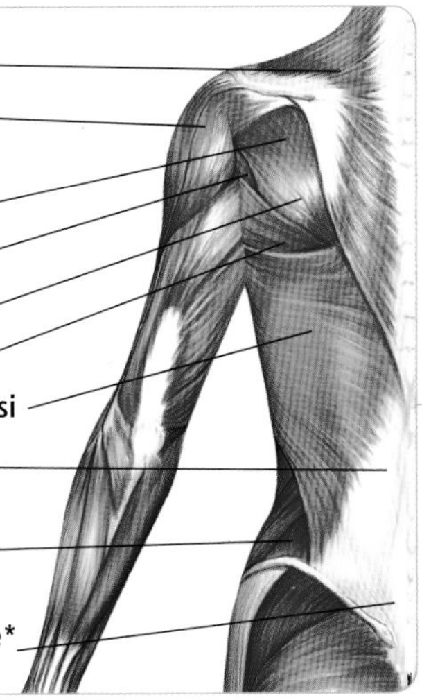

trapezius
deltoideus medialis
infraspinatus
teres minor
subscapularis
teres major
latissimus dorsi
multifidus spinae*
quadratus lumborum
erector spinae*

DO IT RIGHT
- Lift out of your chest and back, rather than depend too much on your arms to create the arch in your back.
- Keep your shoulders and elbows pressed back to create more lift in your chest.

BEST FOR

- quadratus lumborum
- erector spinae
- latissimus dorsi
- gluteus maximus
- gluteus medius
- pectoralis major
- rectus abdominis
- deltoideus
- teres major
- teres minor

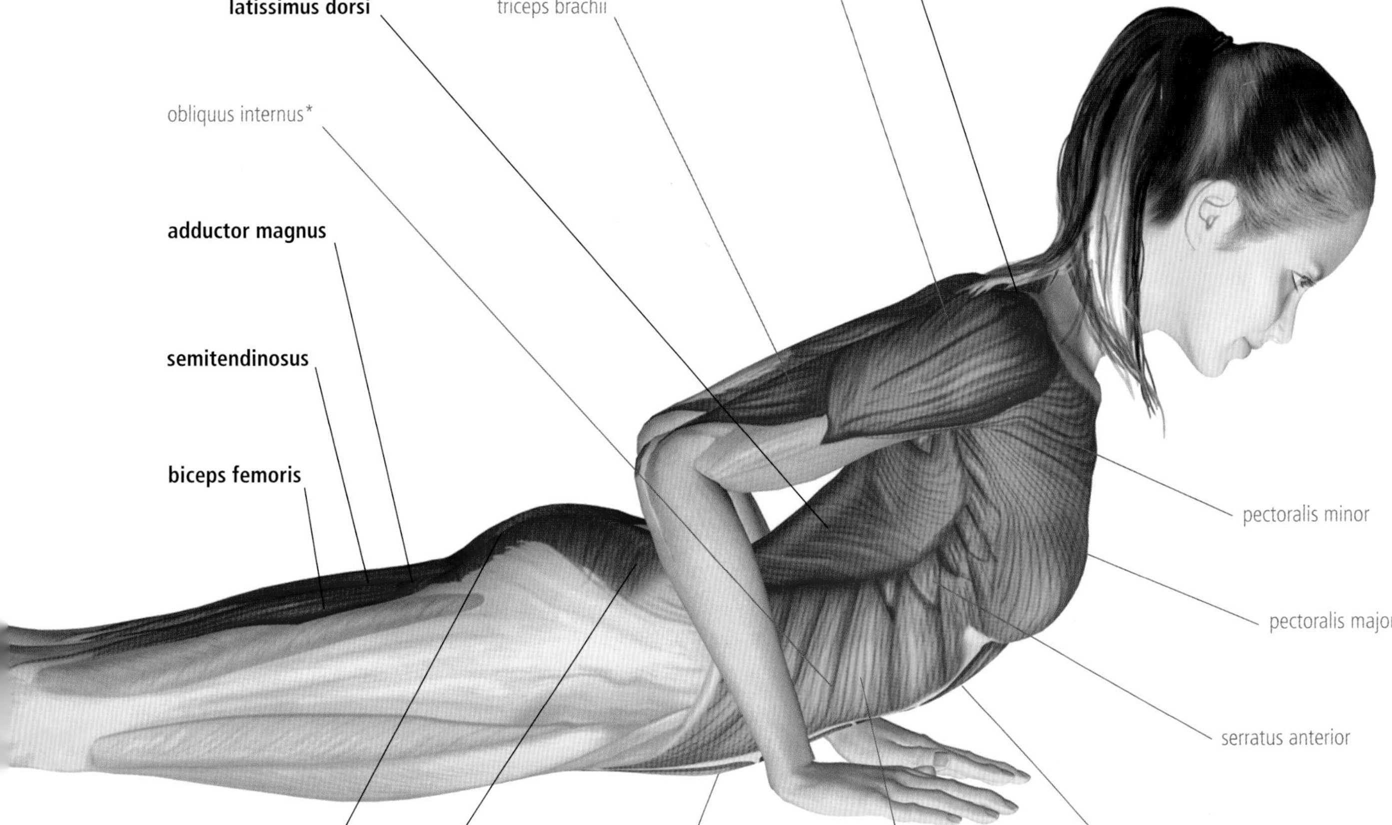

trapezius
deltoideus medialis
latissimus dorsi
triceps brachii
obliquus internus*
adductor magnus
semitendinosus
biceps femoris
pectoralis minor
pectoralis major
serratus anterior
rectus abdominis
gluteus maximus
gluteus medius*
transversus abdominis*
obliquus externus

ANNOTATION KEY
Black text indicates strengthening muscles
Gray text indicates stretching muscles
* indicates deep muscles

HALF-FROG POSE
(ARDHA BHEKASANA)

1 Lie prone on the floor with your legs fully extended. Bend your elbows, placing your hands flat on the floor on either side of your chest. Keep your elbows pulled in toward your body.

2 Inhale, and press into the floor with your hands, lifting your chest and upper torso off the floor. Push your shoulders down and back. Keep your pubis pressed against the floor. Your hands should be placed slightly in front of your torso.

a

3 Bend your left knee, drawing your left heel toward your left buttock. Shift your weight onto your right hand, and reach behind you with your left hand to grasp the inside of your left foot. Continue to lift your chest and push down with your right shoulder.

4 Bend your left elbow up toward the ceiling, and rotate your hand so that it rests on top of your foot with your fingers facing forward. Exhale, and press down on your foot with your left hand to stretch it toward your left buttock.

5 Without separating your legs more than one hip-width apart, deepen the stretch by moving your left foot slightly to the outside of your left thigh, aiming the sole of your foot toward the floor.

6 Hold for 30 seconds to 2 minutes. Repeat on the opposite side.

b

PRONUNCIATION & MEANING
- Ardha Bhekasana (are-dah BEK-has-anna)
- *ardha* = half; *bheka* = frog

LEVEL
- Intermediate

BENEFITS
- Strengthens spine and shoulders
- Stretches chest, abdominals, hip flexors, quadriceps, and ankles

CONTRA-INDICATIONS & CAUTIONS
- High or low blood pressure
- Back injury
- Shoulder injury

DO IT RIGHT
- Keep your hips and shoulders squared forward.
- If you have trouble supporting yourself on your hand, lower yourself onto your forearm and elbow.

AVOID
- Pushing so hard on your foot that it causes discomfort in your knee.
- Sinking into your supporting shoulder.

BEST FOR
- latissimus dorsi
- quadratus lumborum
- erector spinae
- pectoralis major
- deltoideus medialis
- rectus abdominis
- transversus abdominis
- iliopsoas
- vastus intermedius
- rectus femoris
- sartorius
- tibialis anterior
- extensor hallucis

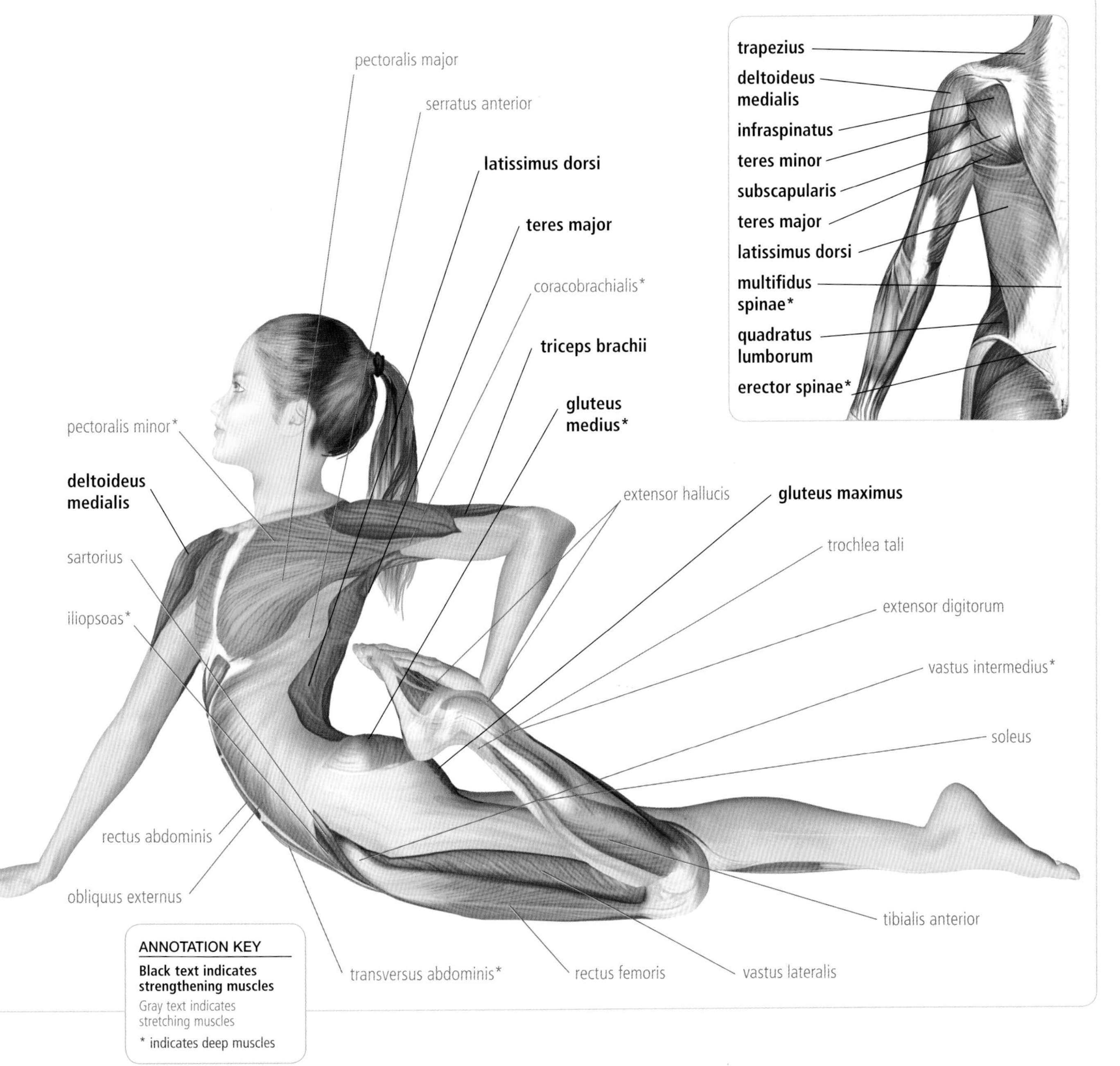

BOW POSE
(DHANURASANA)

❶ Lie prone on the floor, and place your arms by your sides with your palms facing upward.

❷ Place your chin on the floor, and exhale as you bend your knees. Reach your arms behind you, and grasp the outside of your ankles with your hands.

a

❸ Inhale, and lift your chest off the floor. Simultaneously lift your thighs by pulling your ankles up with your hands. Shift your weight onto your abdominals.

❹ Keep your head in a neutral position, and make sure that your knees don't separate more than the width of your hips. Tuck your tailbone into your pubis.

❺ Hold for 20 to 30 seconds. Exhale, and release your ankles, gently returning to the floor.

b

PRONUNCIATION & MEANING
- Dhanurasana (don-your-AHS-anna)
- *dhanu* = bow

LEVEL
- Intermediate

BENEFITS
- Strengthens spine
- Stretches chest, abdominals, hip flexors, and quadriceps
- Stimulates digestion

CONTRA-INDICATIONS & CAUTIONS
- Headache
- High or low blood pressure
- Back injury

DO IT RIGHT
- Keep your knees close together during the duration of the posture, making sure that they don't separate more than the width of your hips.

AVOID
- Holding your breath. Breathing in this pose can be difficult, so make sure to take short, controlled breaths from the back of your torso.
- Rolling back onto your pelvis to support your weight.

BEST FOR

- pectoralis major
- pectoralis minor
- deltoideus
- erector spinae
- gluteus medius
- gluteus maximus
- iliopsoas
- rectus femoris

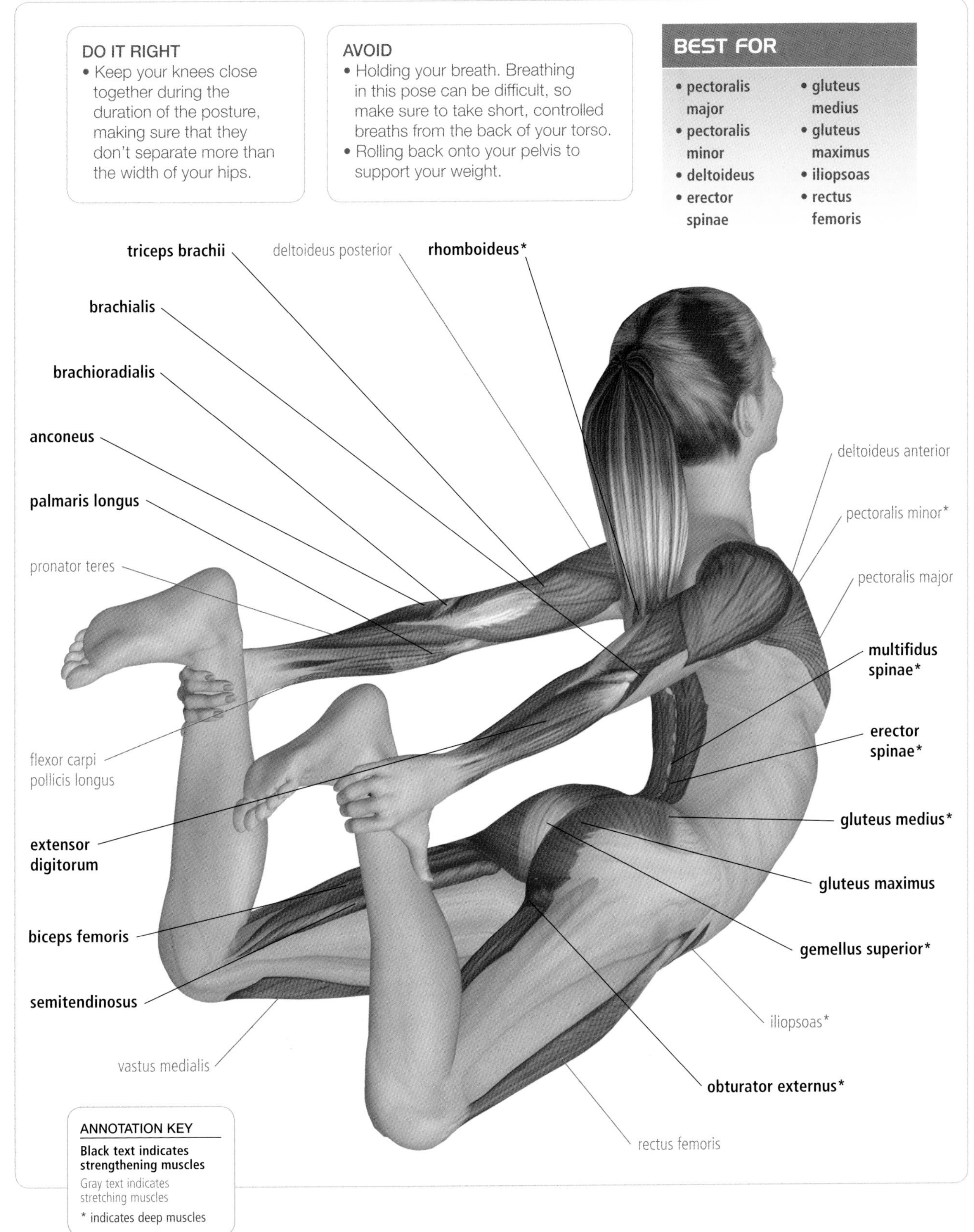

BRIDGE POSE
(SETU BANDHASANA)

ⓐ

❶ Lie supine on the floor. Bend your knees and draw you heels close to your buttocks. Place you hands flat on the floor by your sides.

❷ Exhale, and press down though your feet to lift your buttocks off the floor. With your feet and thighs parallel, push your arms into the floor while extending through your fingertips.

AVOID
- Tucking your chin in toward your chest.
- Using your buttocks more than your hamstrings to lift your hips.

❸ Lengthen your neck away from your shoulders. Lift your hips higher so that your torso rises from the floor.

❹ Hold for 30 seconds to 1 minute. Exhale as you release your spine onto the floor, one vertebra at a time. Repeat at least one more time.

ⓑ

PRONUNCIATION & MEANING
- Setu Bandhasana (SET-too BAHN-dahs-anna)
- *setu* = dam, dike, bridge; *bandha* = lock

LEVEL
- Beginner

BENEFITS
- Strengthens thighs and buttocks
- Stretches chest and spine
- Stimulates digestion
- Stimulates thyroid
- Reduces stress

CONTRA-INDICATIONS & CAUTIONS
- Shoulder injury
- Back injury
- Neck issues

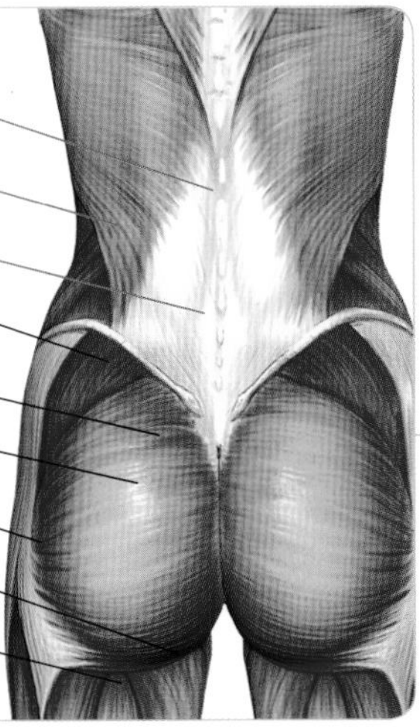

multifidus spinae*
latissimus dorsi
erector spinae*
gluteus medius*
piriformis*
gluteus maximus
quadratus femoris*
obdurator internus*
obdurator externus*

DO IT RIGHT
- Roll your shoulders under once your hips are raised.
- Keep your knees over your heels.
- Tighten your buttocks and your thighs.

BEST FOR

- **sartorius**
- **rectus femoris**
- **iliopsoas**
- **gluteus maximus**
- **gluteus medius**
- **erector spinae**

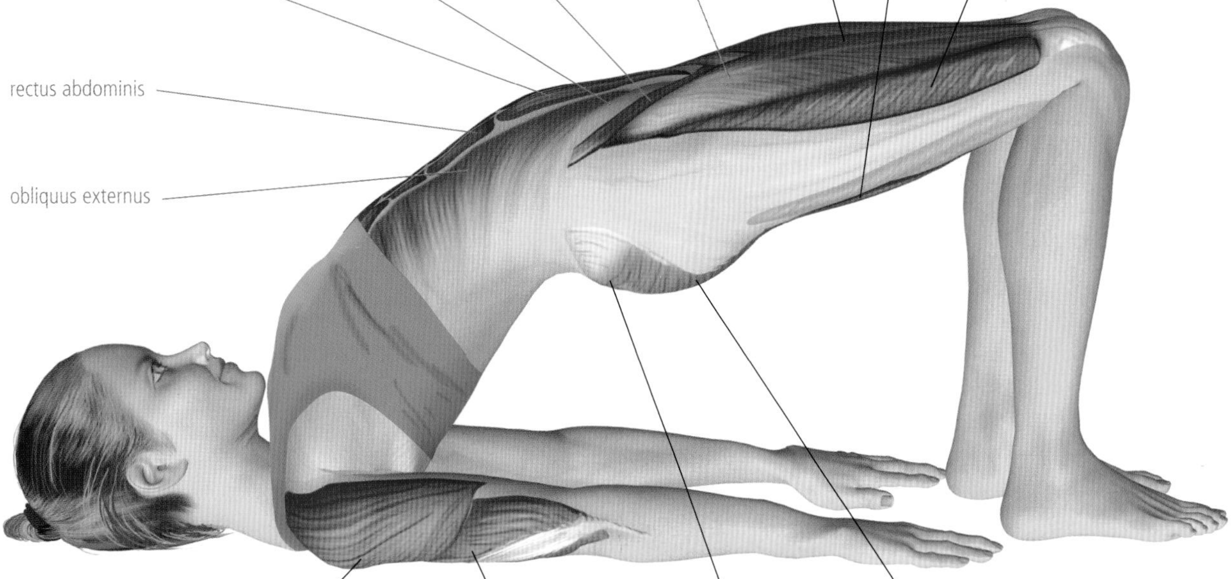

biceps femoris
rectus femoris
vastus lateralis
sartorius
vastus intermedius*
iliopsoas*
transversus abdominis*
rectus abdominis
obliquus externus
deltoideus medialis
triceps brachii
gluteus medius
gluteus maximus

ANNOTATION KEY
Black text indicates strengthening muscles
Gray text indicates stretching muscles
* indicates deep muscles

UPWARD-FACING BOW POSE
(URDHVA DHANURASANA)

ⓐ

❶ Lie supine on the floor. Bend your knees, and draw your heels as close to your buttocks as possible. Bend your elbows, and place your hands on the floor beside your head, with your fingertips pointing toward your shoulders.

ⓑ

❷ Exhale, and push down into your feet to lift your buttocks off the floor. Tighten your thighs, and keep your feet parallel. Push your hands into the floor to raise yourself onto the crown of your head.

❸ After a couple of breaths, exhale, and press into the floor with your hands and feet, lifting your hips up toward the ceiling. Straighten your arms, and allow your head to hang in between your shoulders. Push through your legs, straightening them as much as possible. Open your shoulders, and feel the extension through your entire spine.

ⓒ

❹ Hold for 5 to 30 seconds. Exhale as you bend your arms, and slowly lower yourself to the floor. Repeat at least one more time.

PRONUNCIATION & MEANING
- Urdhva Dhanurasana (OORD-vah don-your-AHS-anna)
- *urdhva* = upward; *dhanu* = bow
- Also called Wheel Pose

LEVEL
- Intermediate/ Advanced

BENEFITS
- Strengthens thighs and buttocks
- Stretches chest and spine
- Stimulates digestion
- Stimulates thyroid
- Reduces stress

CONTRA-INDICATIONS & CAUTIONS
- Back injury
- Carpal tunnel syndrome
- High or low blood pressure
- Headache

DO IT RIGHT

- Lift up, and extend through your shoulders, spine, and quadriceps, being careful not to put all the extension on your lower back.
- Keep your knees close together during the duration of the posture, making sure that they don't separate more than the width of your hips.

BEST FOR

- **deltoideus medialis**
- serratus anterior
- **infraspinatus**
- **rhomboideus**
- **flexor carpi radialis**
- **latissimus dorsi**
- **trapezius**
- **erector spinae**
- **gluteus maximus**
- **vastus lateralis**
- **teres major**
- **teres minor**

AVOID

- Turning your feet out.
- Splaying your elbows out to the sides to push up into the pose.

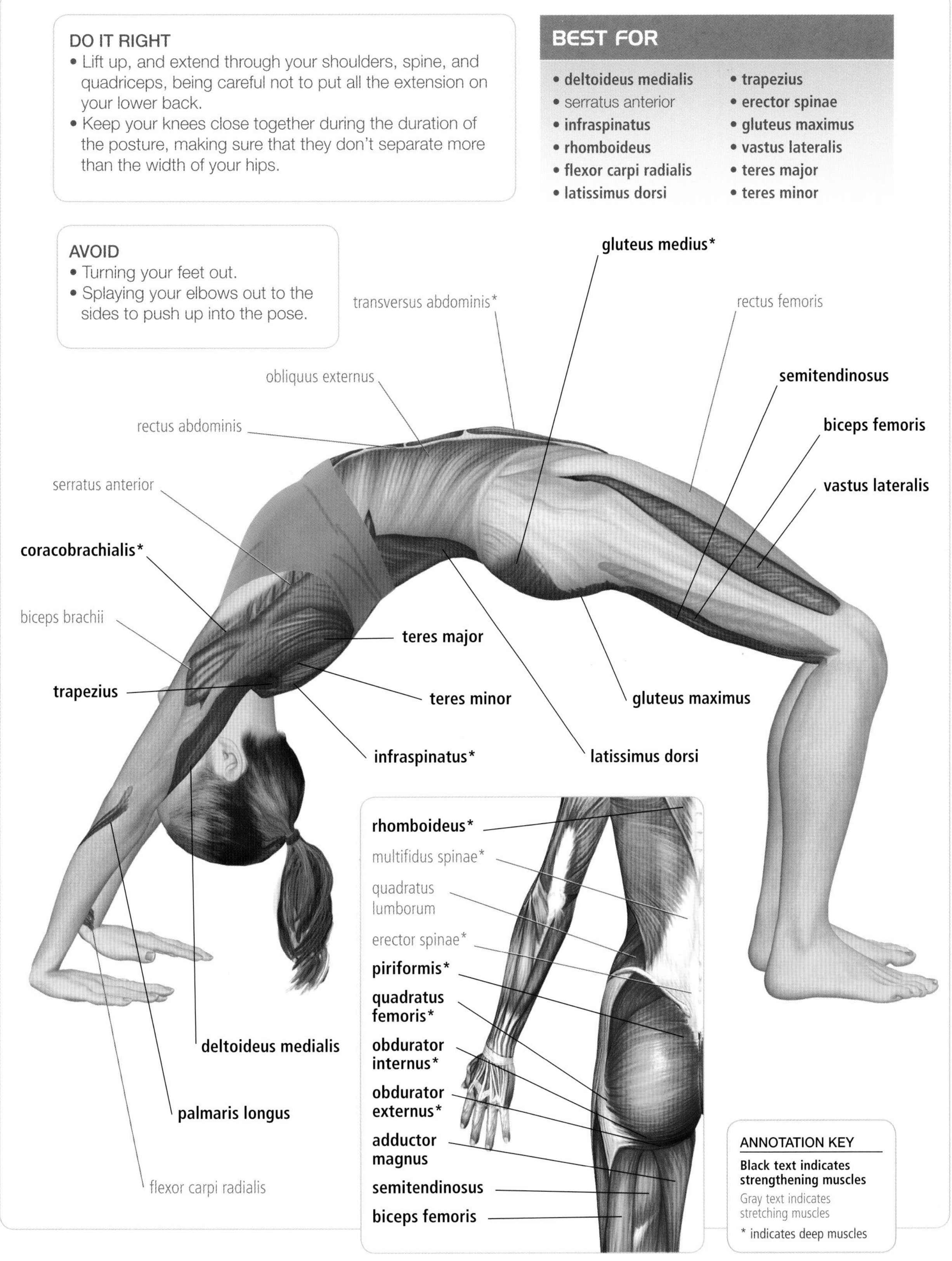

ANNOTATION KEY

Black text indicates strengthening muscles

Gray text indicates stretching muscles

* indicates deep muscles

CAMEL POSE

(UTRASANA)

a

❶ With your knees one hip-width apart, kneel on the floor with your thighs perpendicular to the floor and your hips open. Tuck your tailbone toward your pubis, and lift up through your spine.

b

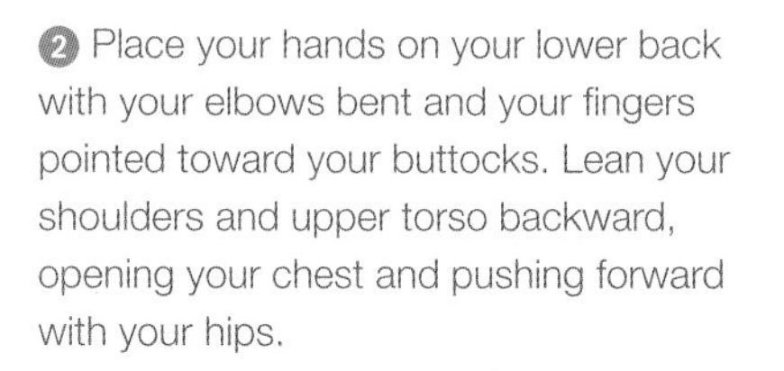

❷ Place your hands on your lower back with your elbows bent and your fingers pointed toward your buttocks. Lean your shoulders and upper torso backward, opening your chest and pushing forward with your hips.

❸ Exhale, and drop back, pressing your pelvis upward and elongating your spine. Pressing your shoulder blades back, lean slightly to your right, and place your right hand on your right heel. Lean slightly to your left, and place your left hand on your left heel. Your fingers should be pointed toward your toes.

❹ Push your thighs forward and center your weight in between your knees, lifting your chest into the arch. Drop your head back, and relax your throat.

❺ Hold for 20 seconds to 1 minute. To come out of the pose, contract your abdominals to lift your chest forward, and slowly bring your hands to your lower back before returning to the starting position.

c

PRONUNCIATION & MEANING
- Utrasana (oosh-TRAHS-anna)
- *ustra* = camel

LEVEL
- Intermediate

BENEFITS
- Strengthens spine
- Stretches thighs, hip flexors, chest, and abdominals
- Stimulates digestion

CONTRA-INDICATIONS & CAUTIONS
- Back injury
- High or low blood pressure
- Headache

AVOID
- Compressing your lower back.
- Rushing into the back bend, which can strain your back.

DO IT RIGHT
- Keep your pelvis pressed forward, and lift up with your abdominals.

BEST FOR

- pectoralis major
- pectoralis minor
- sternocleidomastoideus
- trapezius
- rectus abdominis
- erector spinae
- gluteus medius
- gluteus maximus
- iliopsoas
- deltoideus anterior
- quadratus lumborum

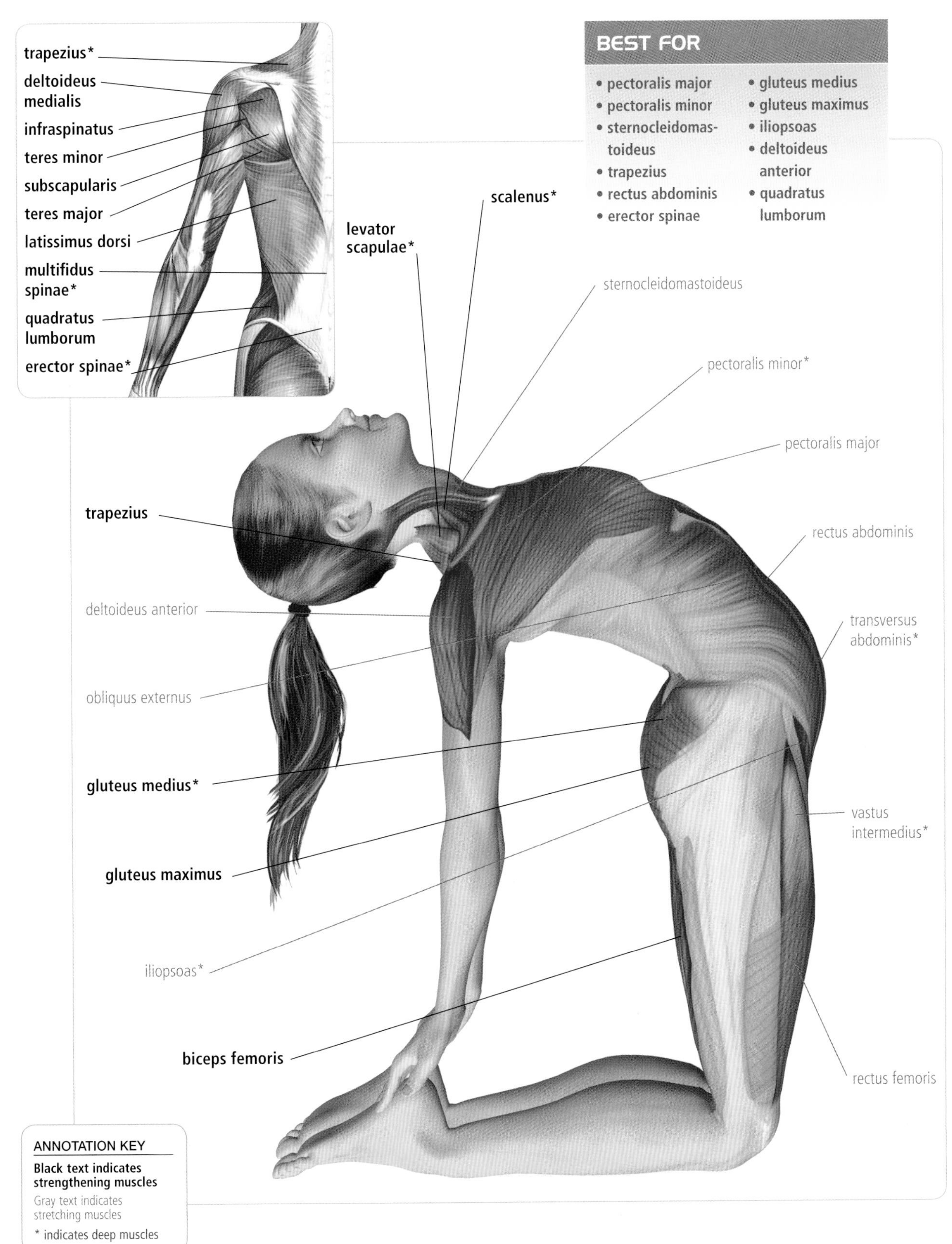

ANNOTATION KEY
Black text indicates strengthening muscles
Gray text indicates stretching muscles
* indicates deep muscles

FISH POSE
(MATSYASANA)

❶ Lie supine on the floor with your arms by your sides. Push down into your heels to lift your hips, and place your hands beneath your buttocks, with your palms facing down.

AVOID
- Pushing your weight onto your head and neck.
- Lifting your hips as you push up into the arch.

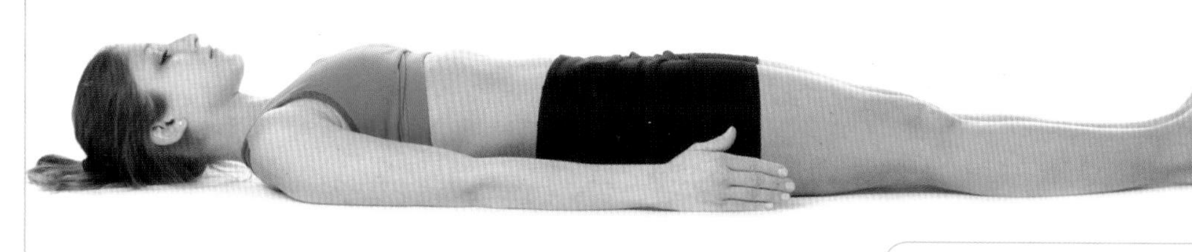

a

DO IT RIGHT
- Keep your elbows and forearms pulled in toward your torso throughout the pose.
- Perform this pose with your legs straight, bent, or in Full Lotus (Padmasana, see page 109).

❷ Rest your buttocks on the tops of your hands, and elongate your legs. Inhale, and press down with your forearms, slightly bending your elbows. Lift up with your chest and head off the floor, creating an arch in your upper back.

❸ Tilt your head back, and place it on the floor. Keep the majority of your weight on your elbows.

❹ Hold for 15 to 30 seconds.

PRONUNCIATION & MEANING
- Matsyasana (mot-see-AHS-anna)
- *matsya* = fish

LEVEL
- Beginner/ Intermediate

BENEFITS
- Stretches chest and abdominals
- Strengthens neck, shoulders, and spine
- Improves posture

CONTRA-INDICATIONS & CAUTIONS
- Back injury
- High or low blood pressure
- Headache

b

c

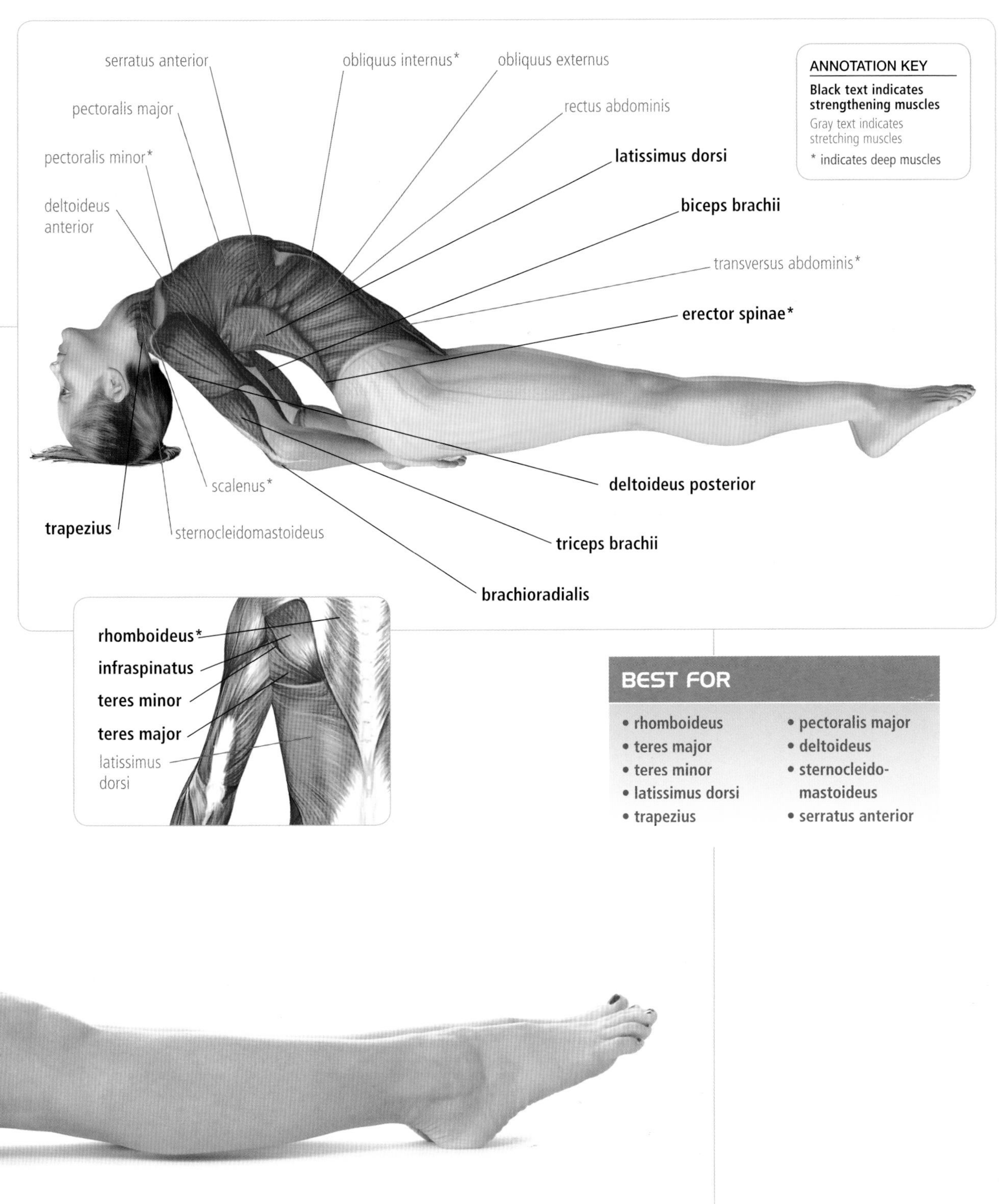

BEST FOR

- rhomboideus
- teres major
- teres minor
- latissimus dorsi
- trapezius
- pectoralis major
- deltoideus
- sternocleido-mastoideus
- serratus anterior

LOCUST POSE
(SALABHASANA)

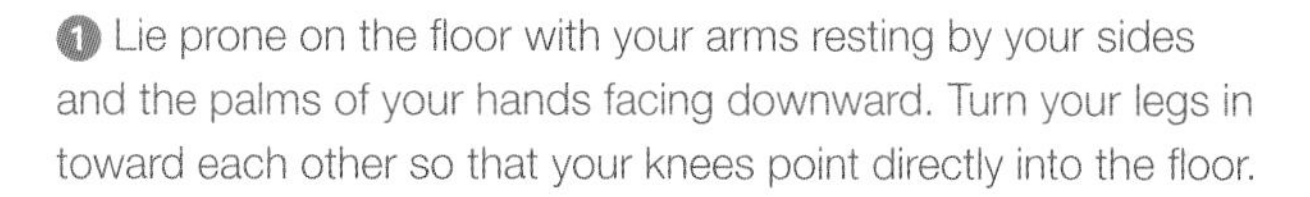

❶ Lie prone on the floor with your arms resting by your sides and the palms of your hands facing downward. Turn your legs in toward each other so that your knees point directly into the floor.

ⓐ

❷ Squeezing your buttocks, inhale, and lift up your head, chest, arms and legs simultaneously. Extend your arms and legs behind you, with your arms parallel to the floor. Lift up as high as possible, with your pelvis and lower abdominals stabilizing your body on the floor. Keep your head in neutral position.

❸ Hold for 30 seconds to 1 minute. Repeat 1 to 2 times.

PRONUNCIATION & MEANING
- Salabhasana (sha-la-BAHS-anna)
- *salabha* = locust, grasshopper

LEVEL
- Beginner

BENEFITS
- Strengthens spine, buttocks, arms, and legs
- Stretches hip flexors, chest, and abdominals
- Stimulates digestion

CONTRA-INDICATIONS & CAUTIONS
- Back injury

AVOID
- Bending your knees.
- Holding your breath.

DO IT RIGHT
- Elongate the back of your neck.
- Open your chest to extend the arch through your entire spine.

ⓑ

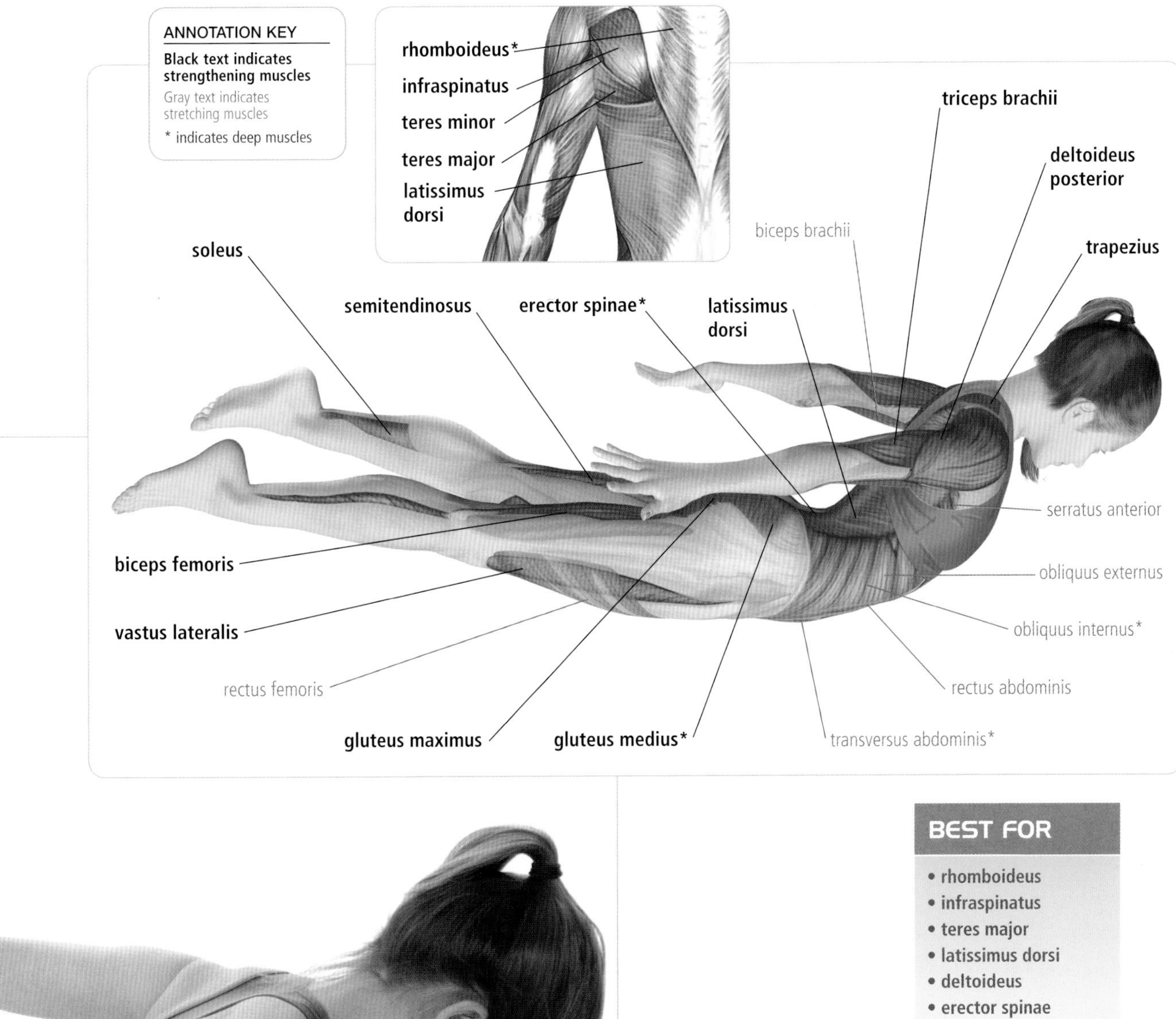

BEST FOR

- rhomboideus
- infraspinatus
- teres major
- latissimus dorsi
- deltoideus
- erector spinae
- trapezius
- gluteus maximus
- gluteus medius

ONE-LEGGED KING PIGEON POSE

(EKA PADA RAJAKAPOTASANA)

BACKBENDS

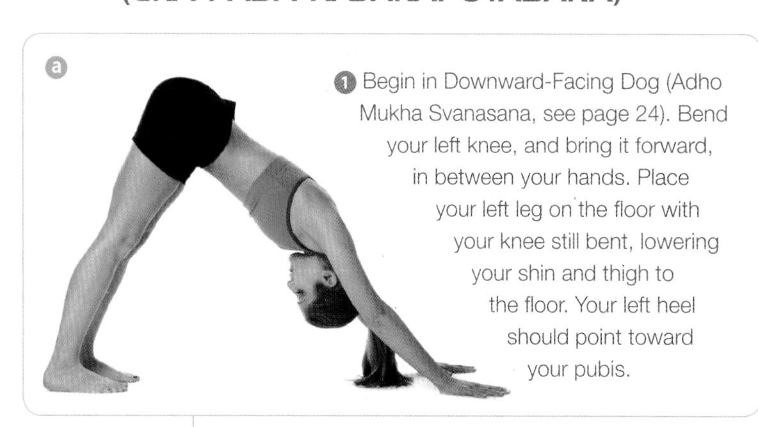

a

1. Begin in Downward-Facing Dog (Adho Mukha Svanasana, see page 24). Bend your left knee, and bring it forward, in between your hands. Place your left leg on the floor with your knee still bent, lowering your shin and thigh to the floor. Your left heel should point toward your pubis.

AVOID

- Compensating for tight shoulders and chest by compressing your lower back.
- Rolling your back knee to either side.

b

2. Extend your right leg behind you. Your hips should be squared forward, and your right knee should point down toward the floor.

3. Lift your chest, using your fingertips to bring your torso to an upright position. Press down into the floor with your hips and pubis, and lift up with your chest.

4. Bend your right knee, and flex your foot, drawing your heel toward your buttock. Reach back with your right hand, your palm facing up, and grasp your toes from the outside of your foot. You may keep your left fingertips on the floor in front of you for balance.

5. Point your right elbow up toward the ceiling, pull your sternum upward, and point your toes. Drop your head back, and reach your left arm over your head to grasp your toes with your left hand. Pull your foot toward your head.

6. Hold for 10 seconds to 1 minute. Return to Downward-Facing Dog, and repeat on the other side.

DO IT RIGHT

- Keep your hips squared forward throughout the pose.
- Sit as deeply as possible into the leg position, drawing your groins toward the floor.

c

PRONUNCIATION & MEANING

- Eka Pada Rajakapotasana (aa-KAH pah-DAH rah-JAH-cop-poh-TAHS-anna)
- *eka* = one; *pada* = foot or leg; *raja* = king; *kapota* = pigeon or dove

LEVEL

- Advanced

BENEFITS

- Stretches hips, thighs, spine, chest, shoulders, neck, and abdominals
- Strengthens spine

CONTRA-INDICATIONS & CAUTIONS

- Hip injury
- Back injury
- Knee injury

MODIFICATIONS

Easier: The One-Legged Pigeon Pose is an advanced pose that requires a great deal of flexibility in your hips, spine, and chest, but you can still benefit from the pose without reaching both hands over your head. With your left leg bent on the floor in front of you and your back thigh pointed down to the floor behind you, bend your right knee. Keep your torso lifted up, and support yourself with the fingertips of your left hand. Point your toes up toward the ceiling. Reach back with your right hand and grasp your foot from the inside of your ankle. Lift up through your spine, open your chest, and hold for 10 seconds to 1 minute. Repeat on the other side.

Easier: Follow steps 1 and 2. Exhale, and slide your arms forward as you fold your torso down until it rests on your left shin.

BEST FOR

- quadratus lumborum
- latissimus dorsi
- sartorius
- vastus intermedius
- iliopsoas
- serratus anterior
- obliquus externus
- pectoralis major
- pectoralis minor
- rectus abdominis

ANNOTATION KEY

Black text indicates strengthening muscles

Gray text indicates stretching muscles

* indicates deep muscles

LORD OF THE DANCE POSE

(NATARAJASANA)

a

1 Standing in Mountain Pose (Tadasana, see page 32), bend your right knee, and draw your right heel toward your buttock. Contract the muscles in your left thigh. Keep both hips open.

2 Turn your right palm outward, reach behind your back, and grasp the inside of your right foot with your hand. Lift through your spine from your tailbone to the top of your neck.

3 Raise your right foot toward the ceiling, and push back against your right hand. Lift your left arm up toward the ceiling simultaneously. It is natural to tilt your torso forward while raising your back leg. Lifting your chest and arm will help you stand upright and increase your flexibility.

4 Hold for 20 seconds to 1 minute. Release your foot and repeat on the other side.

b

AVOID
- Looking down at the floor, causing you to lose your balance.
- Compressing your lower back.

PRONUNCIATION & MEANING
- Natarajasana (not-ah-raj-AHS-anna)
- *nata* = dancer; *raja* = king
- Also called the King of the Dance Pose

LEVEL
- Advanced

BENEFITS
- Stretches thighs, groins, abdominals, shoulders, and chest
- Strengthens spine, thighs, hips, and ankles
- Improves balance

CONTRA-INDICATIONS & CAUTIONS
- Back injury
- Low blood pressure

DO IT RIGHT
- Keep your standing leg straight and your muscles contracted.
- If you have trouble maintaining your balance, practice with your free hand against a wall for support.

BEST FOR

- latissimus dorsi
- pectoralis major
- pectoralis minor
- deltoideus
- iliopsoas
- biceps femoris
- semitendinosus
- quadratus lumborum
- serratus anterior

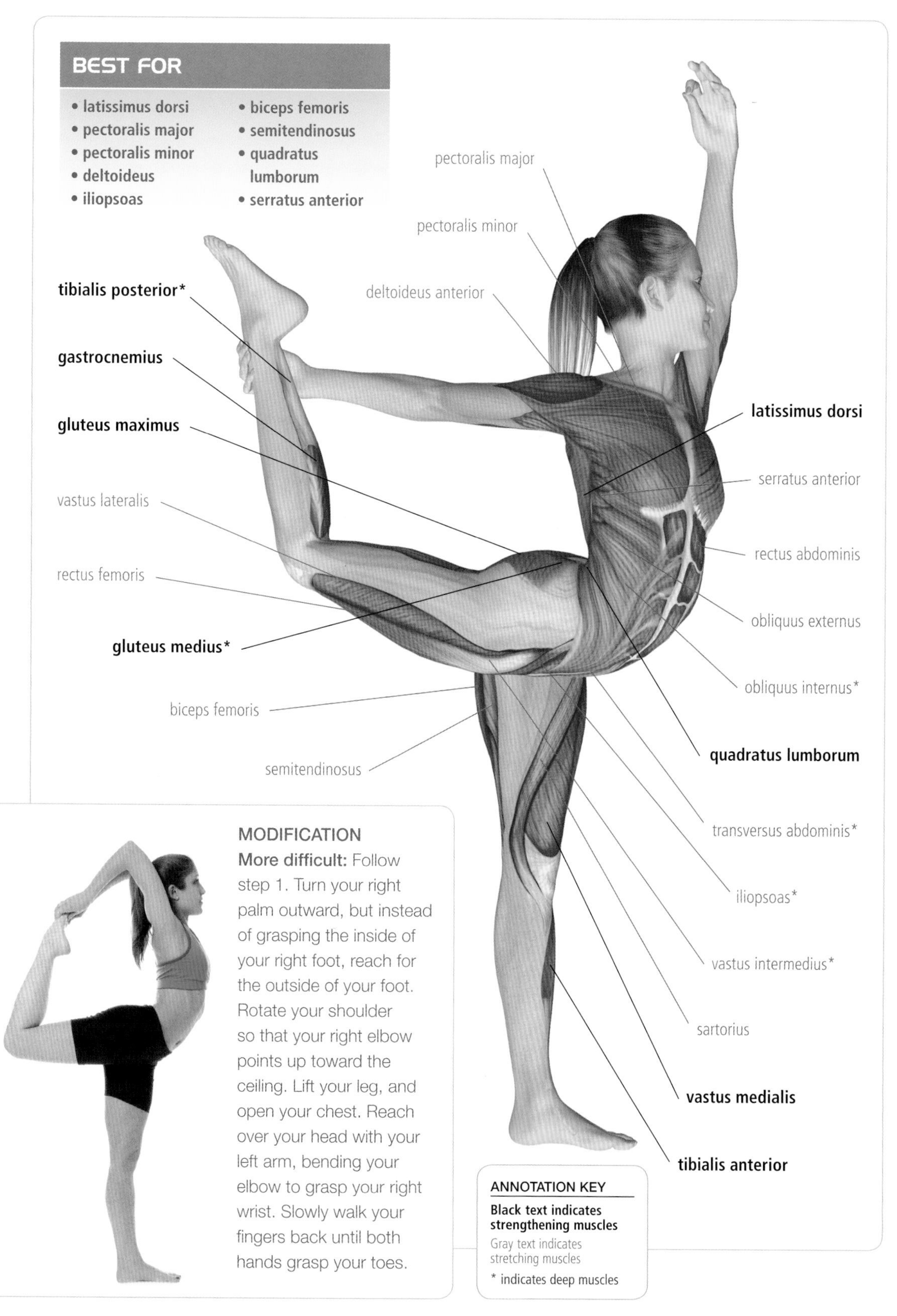

MODIFICATION

More difficult: Follow step 1. Turn your right palm outward, but instead of grasping the inside of your right foot, reach for the outside of your foot. Rotate your shoulder so that your right elbow points up toward the ceiling. Lift your leg, and open your chest. Reach over your head with your left arm, bending your elbow to grasp your right wrist. Slowly walk your fingers back until both hands grasp your toes.

ANNOTATION KEY

Black text indicates strengthening muscles

Gray text indicates stretching muscles

* indicates deep muscles

SEATED POSES & TWISTS

Seated and twisting poses are refreshing poses that counteract the effects of slouching and spinal lethargy. Maintaining the proper alignment in your spine and grounding your sit bones while practicing seated poses will open your hips, groins, pelvis, and lower back. These tend to be the most stable asanas, enabling you to focus on your breath and posture.

Your muscles contract and stretch on opposite sides of your body during twisting poses. These movements target your internal organs and circulatory system, creating a cleansing effect. Your organs are compressed in the pose and refreshed upon release, and internal toxins are purged. Elongating your spine while twisting is crucial, because it will increase your spinal rotation.

HERO POSE
(VIRASANA)

1. Kneel on your hands and knees on the floor. Your thighs should be perpendicular to the floor, and your feet should be angled slightly wider than your hips.

2. Bring your knees together until they touch, pushing the tops of your feet into the floor. Lean forward slightly with your torso, exhaling, and begin to sit back onto your buttocks.

3. Sit on the floor with your buttocks in between your heels.

4. Lift your chest, and press your shoulders back and down, lengthening your tailbone into the floor so that you are resting on your sit bones. Place your hands on the tops of your thighs. Pull your abdominals in toward your spine.

5. Hold for 30 seconds to 1 minute.

AVOID
- Tensing your shoulders up to your ears.
- Turning the soles of your feet out to the sides.
- Sitting on top of your heels.

DO IT RIGHT
- If you experience pain in your knees, place a folded blanket beneath you to elevate your hips. Point your big toes slightly inward so that the tops of your feet lie flat on the floor.

PRONUNCIATION & MEANING
- Virasana (veer-AHS-anna)
- *vira* = man, hero, chief

LEVEL
- Beginner

BENEFITS
- Loosens thighs, knees, and ankles
- Counterbalances hip-opening postures such as the Lotus Pose (Padmasana, see pages 108–109)
- Calms the brain for meditation
- Alleviates high blood pressure

CONTRA-INDICATIONS & CAUTIONS
- Knee injury
- Ankle injury

obliquus internus
rectus abdominis
iliopsoas*
iliacus*
pectineus*
sartorius
vastus intermedius*
vastus lateralis
vastus medialis
tibialis anterior
obliquus externus
transversus abdominis*
tensor fasciae latae
adductor longus
rectus femoris
gracilis*
soleus
gastrocnemius
flexor digitorum
extensor digitorum
extensor hallucis
peroneus
adductor hallucis

ANNOTATION KEY
Black text indicates strengthening muscles
Gray text indicates stretching muscles
* indicates deep muscles

BEST FOR
- rectus femoris
- vastus intermedius
- tensor fasciae latae
- sartorius
- vastus medialis
- vastus lateralis
- tibialis anterior
- extensor hallucis
- peroneus

RECLINING HERO POSE
(SUPTA VIRASANA)

❶ Begin in Hero Pose (Virasana, see page 102). Make sure that you are comfortable sitting with your buttocks completely on the floor.

❷ Lean back gradually, and exhale, putting your hands on the floor behind you for support. Lower yourself onto your elbows.

❸ Recline all the way back until your back reaches the floor. Move your arms to your sides, relaxing them with your palms facing upward. Squeeze your knees together so that they don't separate wider than your hips, and don't allow them to lift off the floor.

❹ Hold for 30 seconds to 1 minute.

BEST FOR

- iliopsoas
- pectineus
- sartorius
- biceps femoris
- vastus intermedius
- vastus medialis
- tibialis anterior
- rectus femoris

DO IT RIGHT

- If you are able to comfortably sit in Hero Pose but experience difficulty reclining to the floor, place folded blankets underneath your back and neck for support.

AVOID

- Sliding your knees beyond the width of your hips.
- Pushing yourself down—relax, and breathe through the reclining action.

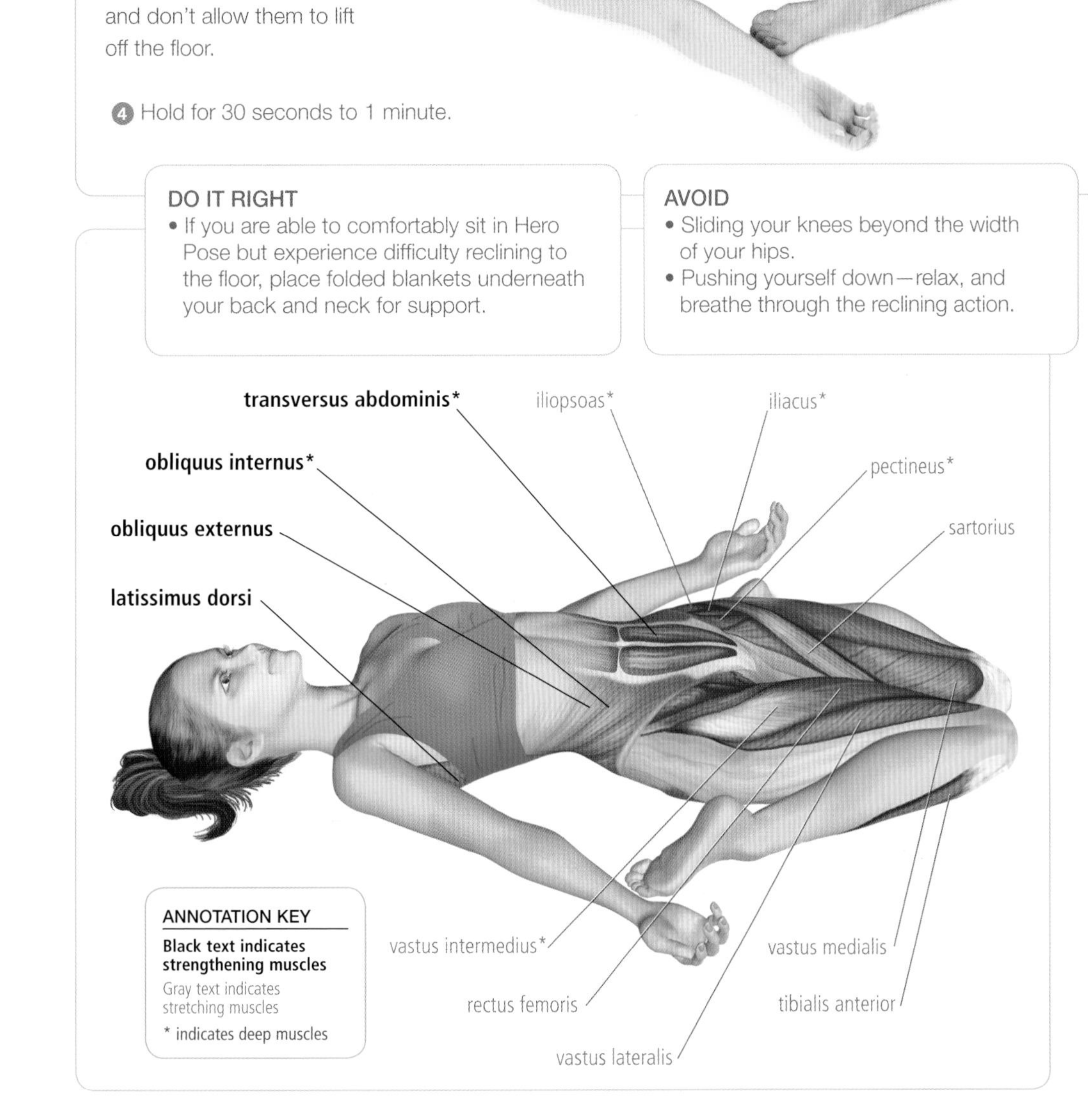

ANNOTATION KEY
Black text indicates strengthening muscles
Gray text indicates stretching muscles
* indicates deep muscles

PRONUNCIATION & MEANING

- Supta Virasana (soup-tah veer-AHS-anna)
- *supta* = reclining, lying down; *vira* = man, hero, chief

LEVEL

- Intermediate

BENEFITS

- Loosens thighs, knees, hip flexors, and ankles
- Stimulates digestion
- Alleviates arthritis
- Alleviates respiratory problems

CONTRA-INDICATIONS & CAUTIONS

- Knee injury
- Ankle injury
- Back issues

BOUND ANGLE POSE

(BADDHA KONASANA)

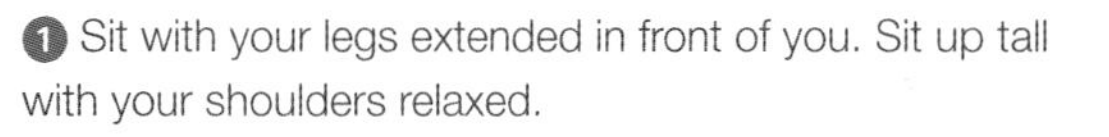

1. Sit with your legs extended in front of you. Sit up tall with your shoulders relaxed.

2. Bring your knees toward your chest with your feet flat on the floor.

3. Exhale, and open hips, drawing your thighs to the floor. Use your hands to press your feet together, and keep the outsides of your feet on the floor.

4. Draw your torso upward, and focus on keeping the spine in the neutral position. Your weight should be balanced evenly on your sit bones. Allow your hips to open farther and your thighs to drop to the floor.

5. Hold for 1 to 5 minutes.

AVOID
- Pushing your knees down with your hands.
- Rounding your back.

BEST FOR
- **iliopsoas**
- **tensor fasciae latae**
- **adductor magnus**
- **adductor longus**
- **iliacus**

DO IT RIGHT
- Lift upward from your spine, and keep your chest and shoulders pressed open, creating a straight line from your sit bones to your shoulders.
- If your groins and inner thighs are very tight, place a folded blanket beneath your buttocks for elevation.
- If you are comfortable in the pose and want to deepen the stretch, bend forward, leading with your chest.

PRONUNCIATION & MEANING
- Baddha Konasana (BAH-dah cone-AHS-anna)
- *baddha* = bound; *kona* = angle
- Also called Tailor Pose

LEVEL
- Beginner

BENEFITS
- Stretches inner thighs, groins, and knees
- Provides relief from menstrual discomfort

CONTRA-INDICATIONS & CAUTIONS
- Knee injury
- Groin injury

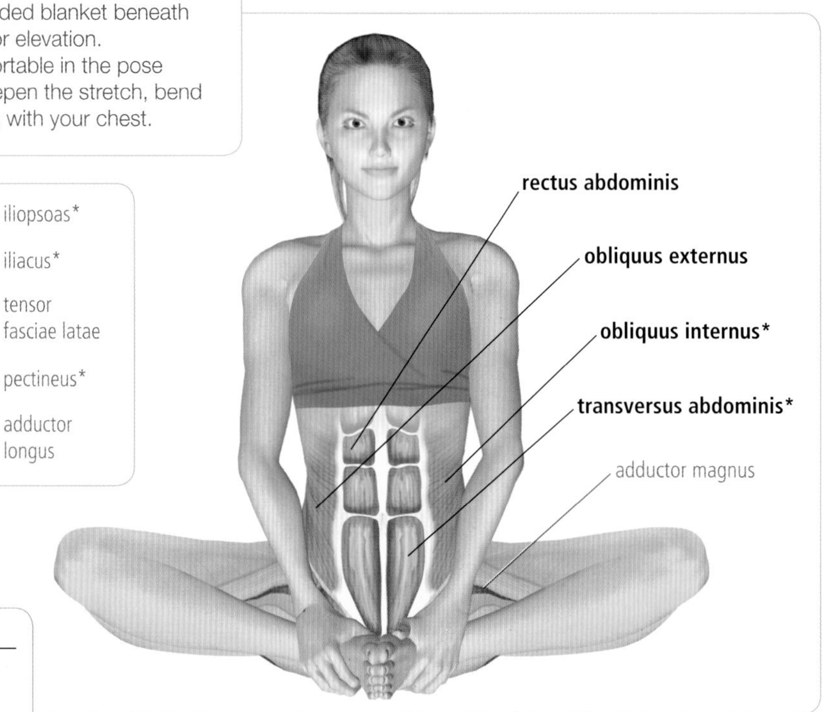

ANNOTATION KEY

Black text indicates strengthening muscles

Gray text indicates stretching muscles

* indicates deep muscles

FIRE LOG POSE
(AGNISTAMBHASANA)

❶ Sit in Easy Pose (Sukhasana, see page 22) with your torso lifted tall.

❷ Place your right ankle on top of your left knee. Your right foot should rest on the outside of your left knee.

❸ Slide your left ankle below your right knee, so that your shins are stacked one on top of the other. Flex both of your feet.

❹ Lift up in your spine through your torso to sit tall on your sit bones. Exhale, and allow your hips to stretch open.

❺ Hold for 1 to 3 minutes. Uncross your legs, and repeat with your left leg on top.

AVOID
- Allowing your feet and ankles to cave inward.

SEATED POSES & TWISTS

PRONUNCIATION & MEANING
- Agnistambhasana (AHG-nih stom-BAHS-anna)
- *agni* = fire; *stambha* = pilla

LEVEL
- Intermediate

BENEFITS
- Stretches hips and groins

CONTRA-INDICATIONS & CAUTIONS
- Knee injury
- Groin injury

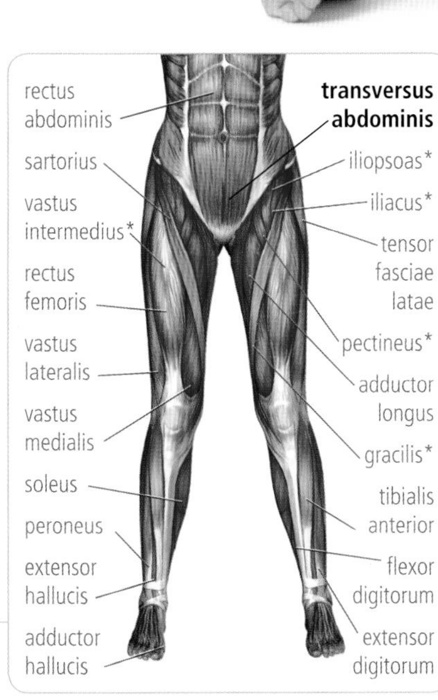

BEST FOR
- **iliopsoas**
- **iliacus**
- **adductor magnus**
- **adductor longus**
- **tensor fasciae latae**
- **pectineus**
- **vastus lateralis**
- **iliacus**
- **vastus medialis**
- **gracilis**
- **sartorius**

DO IT RIGHT
- Rotate out from your hips, rather than from your knees.
- If you experience discomfort when bringing your bottom ankle below your top knee, keep your foot tucked toward your back hip and focus on the position of your top ankle.

COW-FACE POSE

(GOMUKHASANA)

ⓐ

1 Sit in Fire Log Pose (Agnistabbhasana, see page 105), with your right leg stacked on top of your left.

2 Slide your left ankle to the left and your right ankle to the right so that your knees are stacked on top of each other. Your heels should angle toward your hips at approximately the same distance from your hips.

3 Lift up from your spine, sitting with equal weight on your sit bones. Inhale, and reach your right hand to the side, parallel to the floor.

4 Bend your elbow, and rotate your shoulder downward so that the palm of your hand faces behind you. Reach behind your back, palm still up, and draw your elbow into your right side. Continue to rotate your shoulder downward as you reach upward with your hand until your forearm is parallel to your spine. Your right hand should rest in between your shoulder blades.

DO IT RIGHT

- Allow gravity to stretch your hips open.
- Make sure that whichever leg is on top, the opposite elbow is pointed toward the ceiling.
- If you cannot hook your hands behind your back, try using a strap to help you pull your hands closer together.

ⓑ

5 With your next inhalation, reach your left arm up toward the ceiling with your palm facing the back wall. Exhale, and bend your elbow, reaching your left hand down the center of your back.

6 Hook your hands together behind your back. Lift your chest, and pull your abdominals in toward your spine.

7 Hold for approximately 1 minute. Repeat with your left leg stacked on top of your right, and your right elbow pointed toward the ceiling.

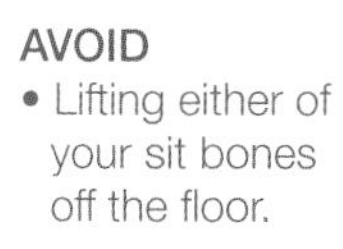

AVOID

- Lifting either of your sit bones off the floor.

PRONUNCIATION & MEANING

- Gomukhasana (go-moo-KAHS-anna)
- *go* = cow; *mukha* = face

LEVEL

- Intermediate

BENEFITS

- Stretches hips, thighs, shoulders, and triceps

CONTRA-INDICATIONS & CAUTIONS

- Shoulder injury

BEST FOR

- deltoideus
- teres minor
- rhomboideus
- subscapularis
- latissimus dorsi
- triceps brachii

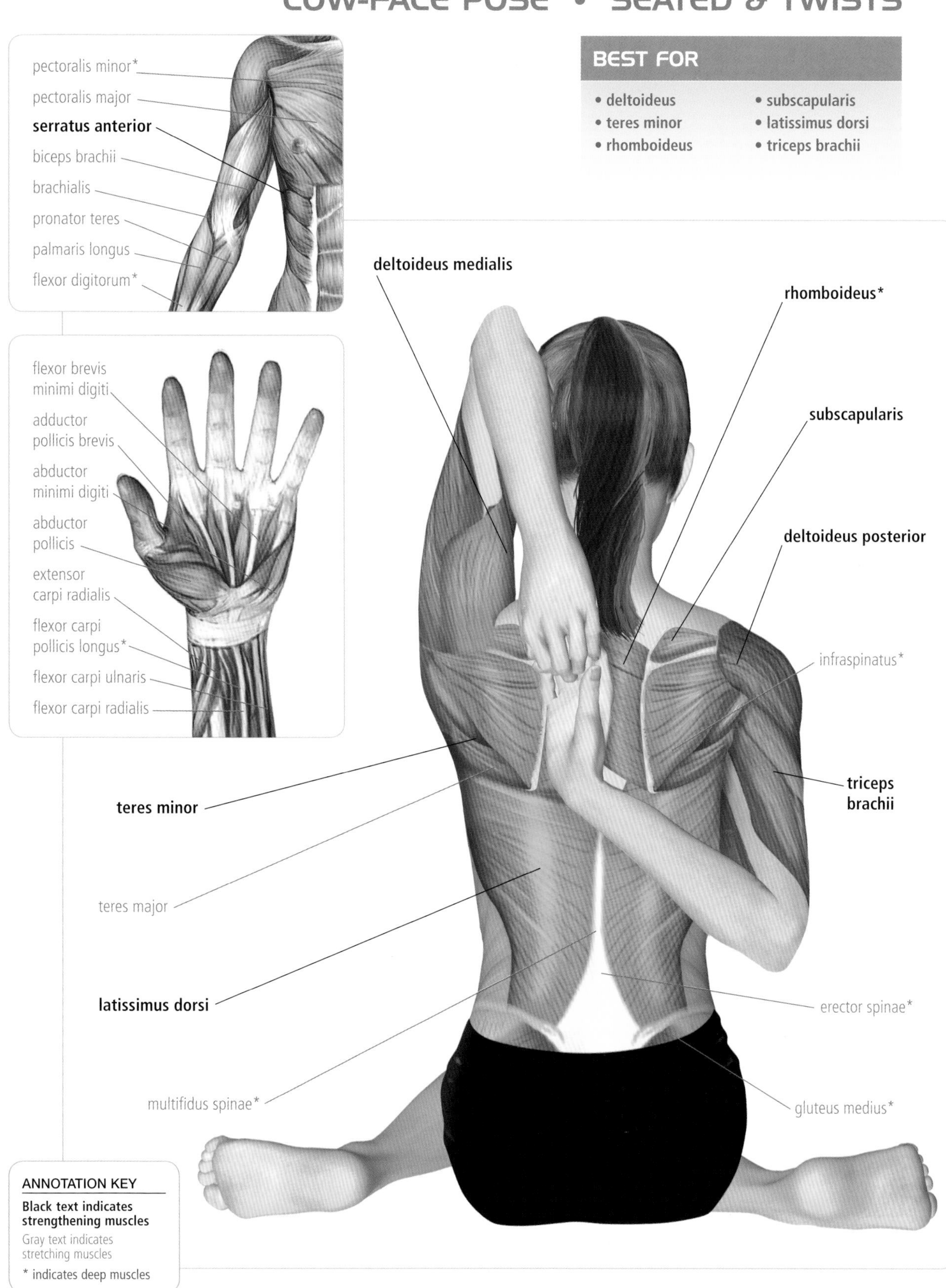

HALF LOTUS POSE
(ARDHA PADMASANA)

1 Sit in Staff Pose (Dandasana, see page 23). Lift up through your spine.

a

BEST FOR
- rectus abdominis
- transversus abdominis
- tibialis anterior
- sartorius
- rectus femoris

b

2 Bend your right knee and open it to the side. Allow your hip to open, and lower your right thigh to the floor.

3 Lean forward slightly, and grab your right shin with your hands. Place your right foot on top of your left thigh, with your heel nestled against your groin. Make sure that the rotation is coming from your hips.

4 Carefully position your left foot beneath your right thigh. Draw your knees closer together. Push into the floor with your groins, as you keep both sit bones on the floor.

5 Extend upward through your spine, and place the backs of your hands on each knee, forming an "O" with your index finger and your thumb.

6 Hold for 5 seconds to 1 minute. Repeat with your left leg on top.

DO IT RIGHT
- Hold the position for the same length of time on both sides.

AVOID
- Overextending your outer ankle.

rectus abdominis
sartorius
vastus intermedius*
rectus femoris
vastus lateralis
vastus medialis
soleus
peroneus
extensor hallucis
adductor hallucis
transversus abdominis
iliopsoas*
iliacus*
tensor fasciae latae
pectineus*
adductor longus
gracilis*
tibialis anterior
flexor digitorum
extensor digitorum

PRONUNCIATION & MEANING
- Ardha Padmasana (are-dah pod-MAHS-anna)
- *ardha* = half; *padma* = lotus

LEVEL
- Intermediate

BENEFITS
- Stretches hips, thighs, and knees, and ankles
- Works the abdominals to stimulate digestion

CONTRA-INDICATIONS & CAUTIONS
- Knee injury

FULL LOTUS POSE
(PADMASANA)

1. Begin in Half Lotus (Ardha Padmasana, see page 108), with your right leg on top of your left.

2. Extend your left leg from below your right hip. With the knee bent, grab your left shin with your hands. Lean back slightly as you bring your left shin on top of your right, and place your left foot on top of your right thigh. Nestle your left heel against your right groin.

3. Push into the floor with your groins and rotate your hips open to press your thighs to the floor. Be sure to keep both sit bones on the floor.

4. Extend upward through your spine and place the backs of your hands on each knee, forming an "O" with your index finger and your thumb.

5. Hold for 5 seconds to 1 minute. Repeat with your right leg on top.

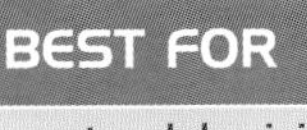

- **rectus abdominis**
- **transversus abdominis**
- **tibialis anterior**

DO IT RIGHT
- If you have trouble keeping your spine in a straight, neutral position, place a folded blanked beneath your hips to elevate your hips above your knees.

AVOID
- Straining your knees. If you experience discomfort in this position, practice the Half Lotus Pose (Ardha Padmasana, see page 108) or the Bound Angle Pose (Baddha Konasana, see page 104) until your hips are flexible enough to practice the Full Lotus.

obliquus externus

rectus abdominis

obliquus internus*

transversus abdominis*

tibialis anterior

ANNOTATION KEY

Black text indicates strengthening muscles

Gray text indicates stretching muscles

* indicates deep muscles

PRONUNCIATION & MEANING
- Padmasana (pod-MAHS-anna)
- padma = lotus

LEVEL
- Advanced

BENEFITS
- Stretches hips, thighs, and knees, and ankles
- Stimulates digestion
- Calms the brain for meditation

CONTRA-INDICATIONS & CAUTIONS
- Knee injury
- Hip injury
- Ankle injury

BOAT POSE
(PARIPURNA NAVASANA)

a

1 Sit on the floor in Staff Pose (Dandasana, see page 23). Lean back slightly, bending your knees, and support yourself with your hands behind your hips. Your fingers should be pointing forward, and your back should be straight.

2 Exhale, and lift your feet off the floor as you lean back from your shoulders. Find your balance point between your sit bones and your tailbone.

AVOID
- Rounding your spine, causing you to sink into your lower back.

b

3 Slowly straighten your legs in front of you so that they form a 45-degree angle with your torso. Point your toes. Lift your arms to your sides, parallel to the floor.

4 Pull your abdominals in toward your spine as they work to keep your balance. Stretch your arms forward through your fingertips, and elongate the back of your neck.

5 Hold for 10 to 20 seconds.

PRONUNCIATION & MEANING
- Paripurna Navasana (par-ee-POOR-nah nah-VAHS-anna)
- *paripurna* = full, entire, complete; *nava* = boat

LEVEL
- Intermediate

BENEFITS
- Strengthens abdominals, hip flexors, spine, and thighs
- Stretches hamstrings
- Stimulates digestion
- Alleviates thyroid problems

CONTRA-INDICATIONS & CAUTIONS
- Neck injury
- Headache
- Lower back pain

BEST FOR

- rectus abdominis
- obliquus internus
- obliquus externus
- iliopsoas
- transversus abdominis
- vastus intermedius
- rectus femoris
- iliacus
- erector spinae

DO IT RIGHT

- Keep your neck elongated and relaxed, minimizing the tension in your upper spine.
- If you are unable to straighten your legs, balance with your knees bent.

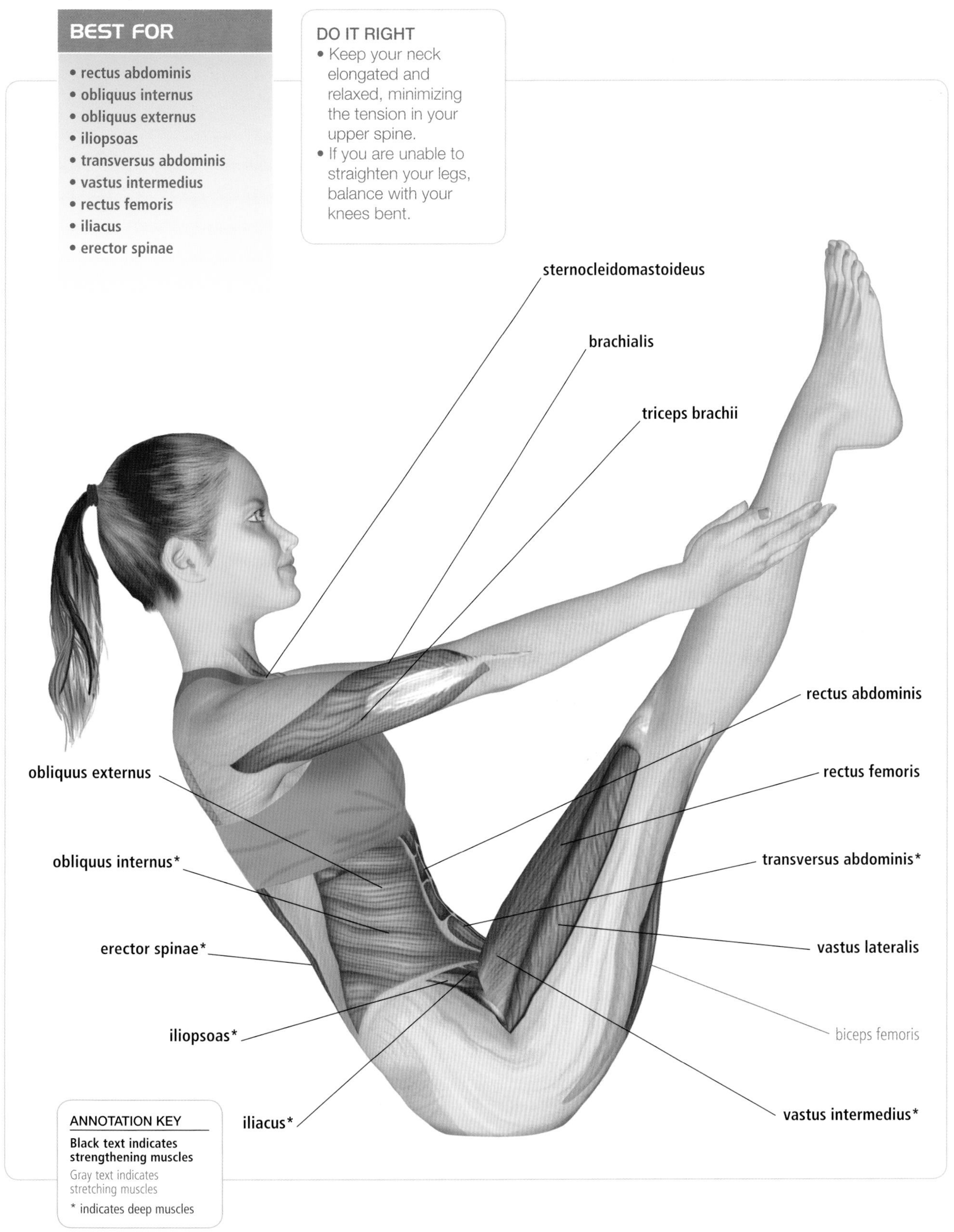

ANNOTATION KEY

Black text indicates strengthening muscles

Gray text indicates stretching muscles

* indicates deep muscles

MONKEY POSE
(HANUMANASANA)

SEATED POSES & TWISTS

❶ Kneel on the floor with your hips open and your back straight.

a

❷ Place your left foot on the floor in front of you to create a lunge position. Make sure that your hips are aligned and facing forward.

❸ Lean forward slightly to balance on your fingertips. Slowly extend your right leg behind you while simultaneously extending your left leg forward.

b

❹ When you have fully lowered yourself to the floor, straighten both legs completely, and point your toes. Your right knee should be facing the floor, and your left knee should face up toward the ceiling. Your hips should be parallel and facing forward.

❺ Lift your chest, and raise your arms toward the ceiling. Arch your back slightly with your shoulders open.

❻ Hold for 30 seconds to 1 minute. Repeat with your right leg in front.

c

DO IT RIGHT
- Practice the Monkey Pose on a hardwood floor or another smooth surface so that you can slide more easily into the pose.
- Push into the floor with your front heel and the top of your back foot as you descend.

PRONUNCIATION & MEANING
- Hanumanasana (ha-new-mahn-AHS-anna)
- *hanuman* = having large jaws; the name of a Hindu deity who appears as a monkey chief

LEVEL
- Advanced

BENEFITS
- Stretches hamstrings and groins

CONTRA-INDICATIONS & CAUTIONS
- Groin or hamstring injury

AVOID
- Pushing yourself too far into the pose—only stretch as far as your hamstrings allow.
- Turning your hips out to the side.

BEST FOR

- iliopsoas
- iliacus
- pectineus
- adductor longus
- sartorius
- vastus intermedius
- rectus femoris
- biceps femoris
- semimembranosus
- semitendinosus

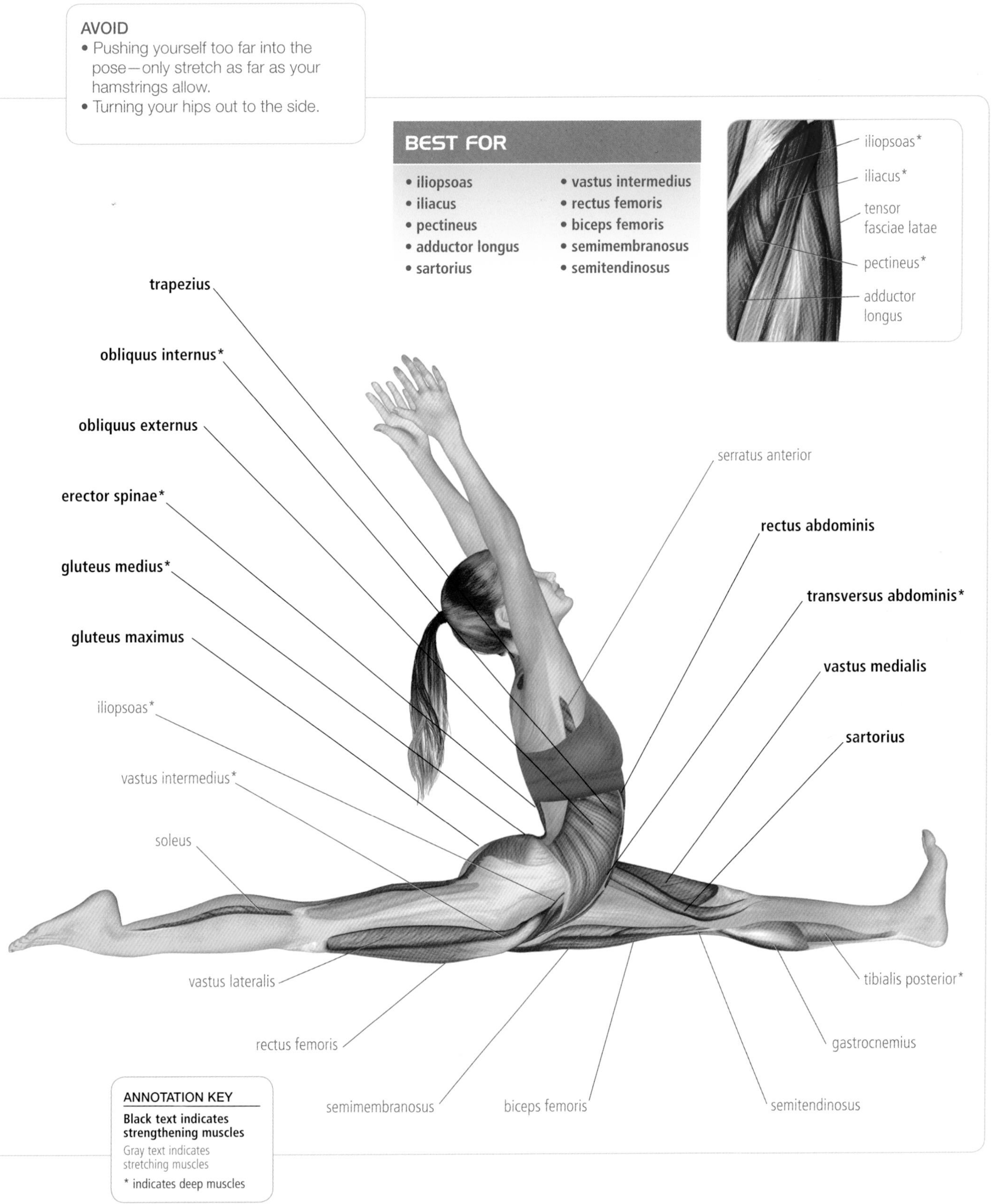

ANNOTATION KEY
Black text indicates strengthening muscles
Gray text indicates stretching muscles
* indicates deep muscles

BHARADVAJA'S TWIST

(BHARADVAJASANA I)

❶ Sit on the floor in Staff Pose (Dandasana, see page 23).

❷ Shift your weight onto your right buttock, and bend your knees to the left, allowing your right thigh to rest on the floor. With your toes pointed toward your left hip, your left thigh should rest on top of your right calf, and your left ankle should sit on top of your right foot.

❸ Inhale, and lift up from your spine. Exhale, and twist to your right, looking over your right shoulder. Place your left hand near your right knee and your right hand on the floor beside your right hip.

❹ With each exhale, deepen the twist while keeping your torso upright and your shoulders pressed back. If possible, bend your right elbow, and reach across your back. Hook your right hand beneath the bend in your left elbow.

❺ Hold for 30 seconds to 1 minute. Repeat on the opposite side.

PRONUNCIATION & MEANING
- Bharadvajasana I (bah-ROD-va-JAHS-anna)
- Bharadvaja = the name of a great Hindu sage

LEVEL
- Beginner

BENEFITS
- Stretches spine, shoulders, and hips
- Stimulates digestion
- Relieves stress

CONTRA-INDICATIONS & CAUTIONS
- Low or high blood pressure
- Diarrhea

AVOID
- Popping your rib cage out.
- Dropping your head.

DO IT RIGHT
- Try to press both sit bones into the floor while twisting.

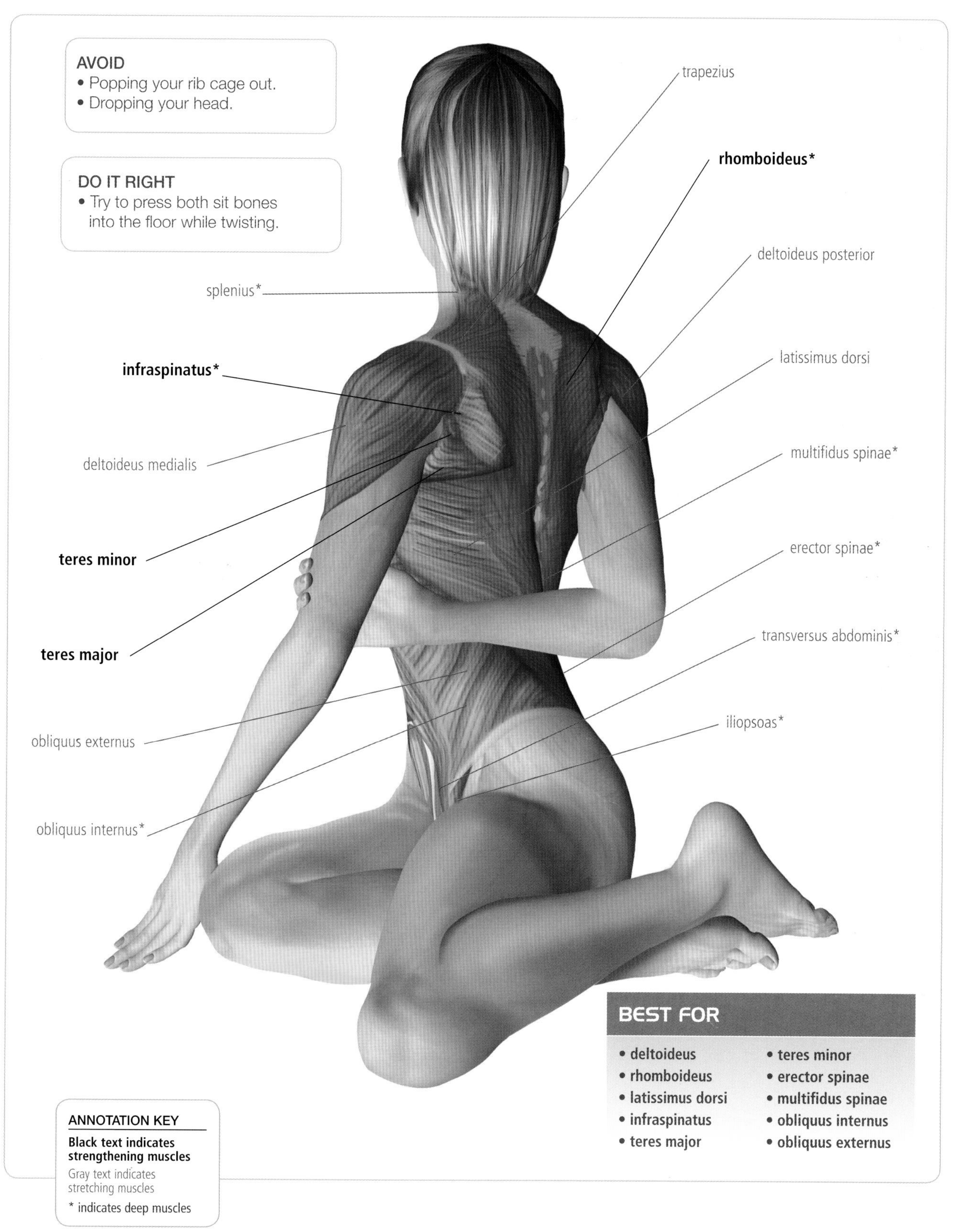

BEST FOR
- deltoideus
- rhomboideus
- latissimus dorsi
- infraspinatus
- teres major
- teres minor
- erector spinae
- multifidus spinae
- obliquus internus
- obliquus externus

ANNOTATION KEY
Black text indicates strengthening muscles
Gray text indicates stretching muscles
* indicates deep muscles

RECLINING TWIST

1 Lie on the floor in Corpse Pose (Savasana, see page 29). Bend your knees with your feet flat on the floor. Extend your arms straight out to the sides, palms facing up.

2 Inhale, and elongate your spine from your hips to the top of your neck. Lift your hips up slightly, and place them on the floor closer to your heels to lengthen and relax your spine further.

3 Lift your feet off the floor, keeping your knees bent.

a

4 Exhale, and bend your knees to the left, causing your hips and spine to twist. Keep your shoulder blades planted on the floor, and allow gravity to pull your left thigh to the floor with each exhalation. Turn your head to the right.

b

4 Hold for 30 seconds to 3 minutes. Repeat on the opposite side.

c

PRONUNCIATION & MEANING
- There is no agreed-upon Sanskrit name for this pose.

LEVEL
- Beginner

BENEFITS
- Releases spinal tension
- Loosens hips
- Tones abdominals

CONTRA-INDICATIONS & CAUTIONS
- Shoulder issues

DO IT RIGHT
- Keep your chest open.
- If you struggle to bring your knees to the floor, place a folded blanket beneath them.
- Experiment with turning your head to both sides. This will change the sensation of the stretch.
- Relax—don't push—into the stretch.

AVOID
- Tensing your shoulders up to your ears.
- Allowing your shoulder blades to lift off the floor. If your shoulder comes up, bend the arm of the lifted shoulder, and place your hand beneath your ribs for support.

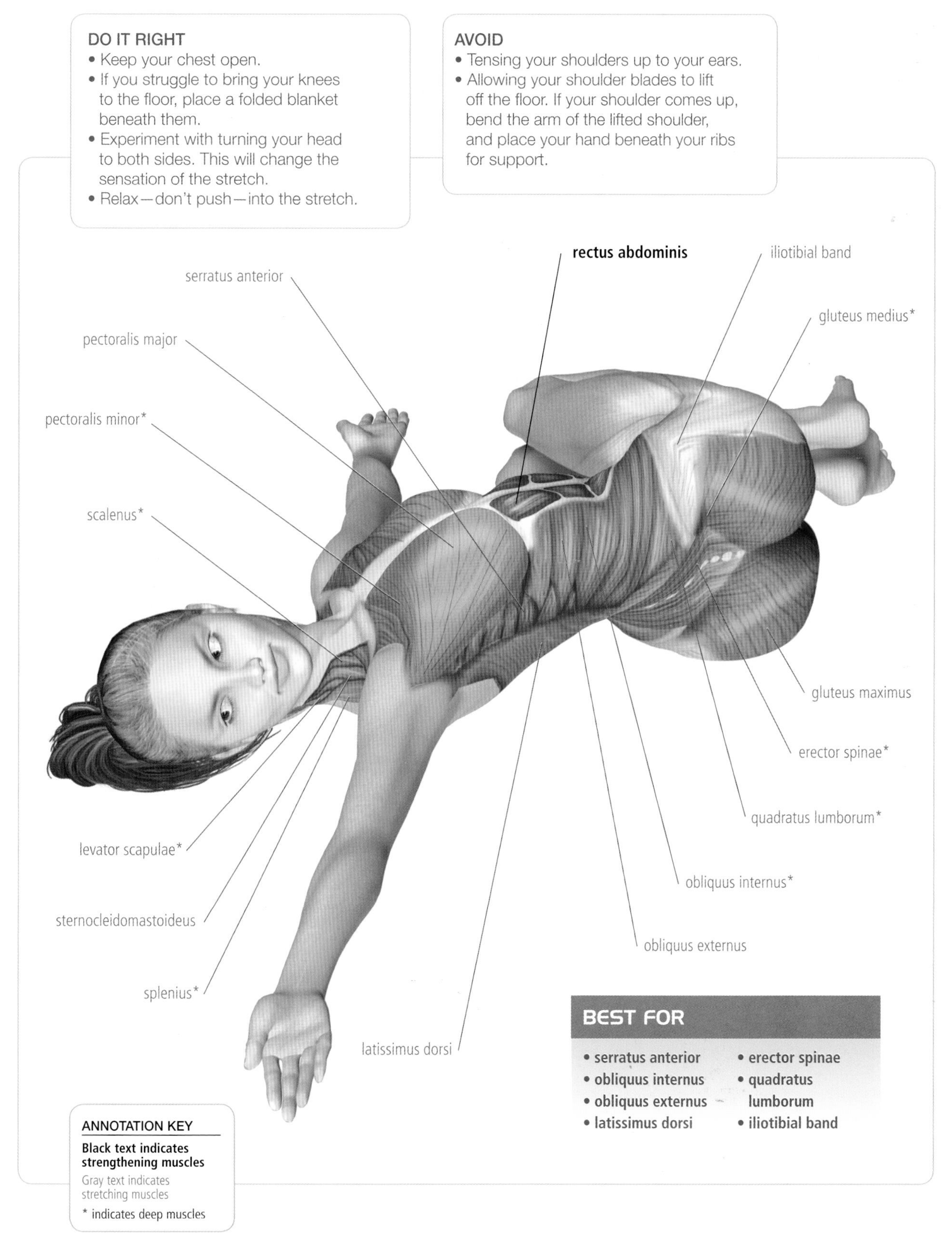

BEST FOR
- serratus anterior
- obliquus internus
- obliquus externus
- latissimus dorsi
- erector spinae
- quadratus lumborum
- iliotibial band

ANNOTATION KEY
Black text indicates strengthening muscles
Gray text indicates stretching muscles
* indicates deep muscles

REVOLVED HEAD-TO-KNEE POSE

(PARIVRTTA JANU SIRSASANA)

1 Begin in Staff Pose (Dandasana, see page 23). Separate your legs wide. Bend your left knee, and draw your heel toward your groin, placing the sole of your foot on your right inner thigh. Lower your left knee to the floor. Draw both sit bones to the floor.

a

DO IT RIGHT
- Elongate the back of your neck.
- Open your chest to extend the arch through your entire spine.

2 Inhale, and lift up through your spine. Exhale, and stretch toward your right leg. Flex your foot, and contract the muscles in your right thigh to push the back of your leg toward the floor. Make sure that your knee is pointed up toward the ceiling.

AVOID
- Bending your extended knees.
- Holding your breath.

b

3 Gently draw your right shoulder to your right inner thigh, and grasp the ball of your foot with your right hand. Keeping your right knee straight, lower your elbow to the floor. Rotate your torso toward the ceiling.

4 Inhale, and stretch your left arm up and over your head to grasp your right foot. Exhale, and press your left shoulder backward to rotate your torso further. Stretch deeper with each exhalation. Gaze up toward the ceiling.

5 Hold for 30 seconds to 1 minute. Repeat with your left leg straight and your right leg bent.

c

PRONUNCIATION & MEANING
- Parivrtta Janu Sirsasana (par-ee-vrit-tah JAH-new shear-SHAHS-anna)
- *parivrtta* = revolved; *janu* = knee; *shiras* = head
- Also called Sitting Leg Stretch

LEVEL
- Intermediate

BENEFITS
- Stretches hamstrings, groins, shoulders and spine
- Stimulates digestion

CONTRA-INDICATIONS & CAUTIONS
- Knee injury
- Shoulder injury

BEST FOR

- gluteus medius
- obliquus internus
- adductor longus
- adductor magnus
- tibialis anterior
- gracilis
- rhomboideus
- trapezius
- latissimus dorsi
- erector spinae
- infraspinatus
- soleus
- gastrocnemius
- semimembranosus
- semitendinosus
- biceps femoris

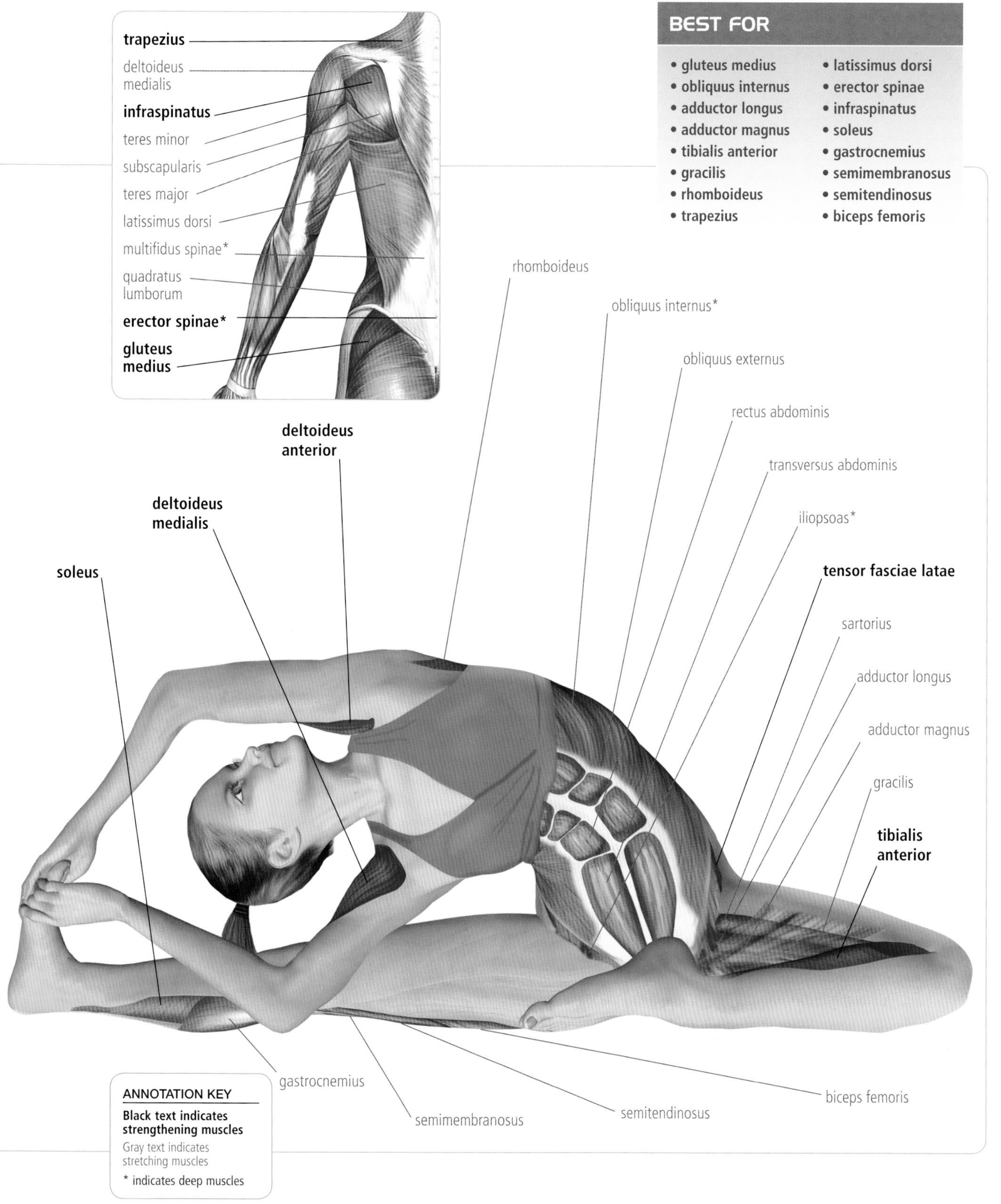

ANNOTATION KEY
Black text indicates strengthening muscles
Gray text indicates stretching muscles
* indicates deep muscles

MARICHI'S POSE

(MARICHYASANA)

1 Sit in Staff Pose (Dandasana, see page 23). Bend your right knee, pulling your heel toward your groin. Keep your left leg extended with your knee pointed up toward the ceiling, and focus on keeping your leg grounded. Place your hands on the floor by your sides.

2 Pushing your right foot and left leg into the floor, inhale, and lift up through your spine and chest. Keep both sit bones on the floor, and relax your shoulders.

3 Exhale, and begin twisting toward your right knee. Wrap your left hand around the outside of your right thigh, pulling your knee in toward your abdominals. Press the fingertips of your right hand on the floor behind your hips. Turn your head to the right.

a

4 Twist deeper with each exhalation. If possible, place your left elbow on the outside of your right knee. Lean back slightly, leading with your upper torso. This will help you twist your entire spine.

5 Hold for 30 seconds to 1 minute. Gently untwist as you exhale, and repeat with your left leg bent and your right elbow over your left knee.

AVOID
- Tensing your shoulders up toward your ears.
- Rounding your spine.
- Forcing a deep twist—gently ease your body into the rotation while maintaining correct posture.

b

PRONUNCIATION & MEANING
- Marichyasana III (mar-ee-chee-AHS-anna)
- *marichi* = ray of light; or name of the Hindu seer credited with intuiting the divine law of the universe, or *dharma*
- Also called the Sage's Pose

LEVEL
- Beginner

BENEFITS
- Stimulates digestion
- Strengthens and stretches spine
- Removes toxins from internal organs

CONTRA-INDICATIONS & CAUTIONS
- High or low blood pressure
- Back injury

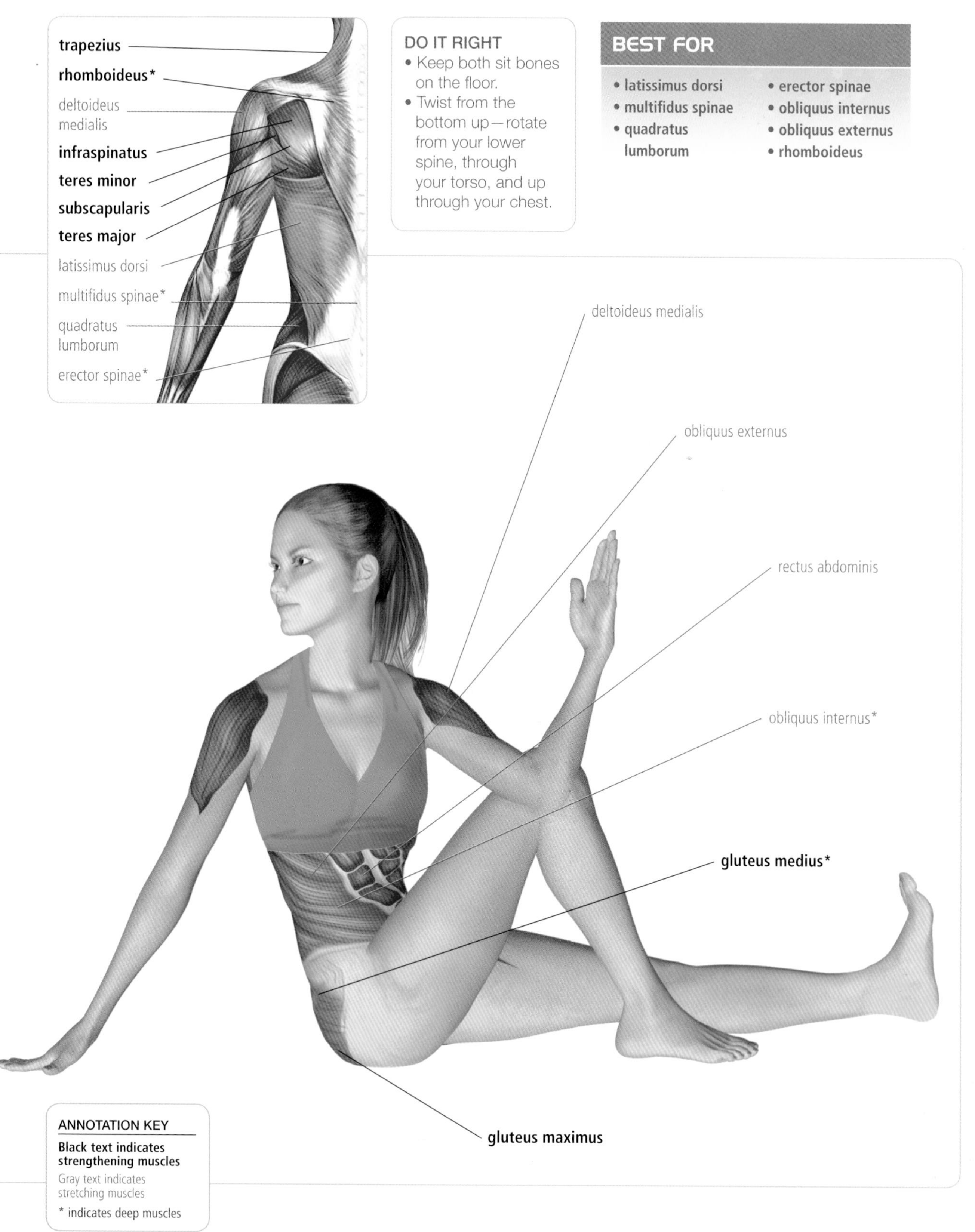

DO IT RIGHT

- Keep both sit bones on the floor.
- Twist from the bottom up—rotate from your lower spine, through your torso, and up through your chest.

BEST FOR

- latissimus dorsi
- multifidus spinae
- quadratus lumborum
- erector spinae
- obliquus internus
- obliquus externus
- rhomboideus

ANNOTATION KEY

Black text indicates strengthening muscles

Gray text indicates stretching muscles

* indicates deep muscles

HALF LORD OF THE FISHES

(ARDHA MATSYENDRASANA)

a

1 Sit in Staff Pose (Dandasana, see page 23). Bend your right knee, and place your right foot over your left leg. Your right foot should be flat on the floor outside of your left thigh.

2 At the same time, bend your left knee, resting the outside of your left thigh on the floor. Your left heel should point toward your right sit bone.

3 Inhale, and lift up through your spine and chest while keeping your shoulders relaxed. Exhale, and begin twisting to your right. Place your left elbow on the outside of your right knee. Press your right hand on the floor behind your hips. Turn your head to the right.

4 Twist deeper with each exhalation. Lean back slightly, leading with your upper torso. Using your left arm, pull your right thigh closer toward your abdominals. Continue to lengthen your spine from the bottom up, pulling your tailbone down toward the floor. Use your right hand to guide your rotation deeper.

5 Hold for 30 seconds to 1 minute. Gently untwist as you exhale, and repeat with your left leg over your right thigh.

b

AVOID
- Tensing your shoulders up toward your ears.
- Rounding your spine.
- Lifting the foot of your raised leg off the ground.

DO IT RIGHT
- Try to pull the thigh of your raised leg and your torso as close together as possible without collapsing your spine.
- Pull your back shoulder toward the back wall as you twist through your entire spine.

PRONUNCIATION & MEANING
- Ardha Matsyendrasana (ARD-ha MOTS-yen-DRAHS-anna)
- *ardha* = half; *matsya* = fish; *indra* = ruler, lord

LEVEL
- Intermediate

BENEFITS
- Stimulates digestion
- Stretches hips, spine, and shoulders
- Relieves backache and menstrual discomfort

CONTRA-INDICATIONS & CAUTIONS
- Back injury

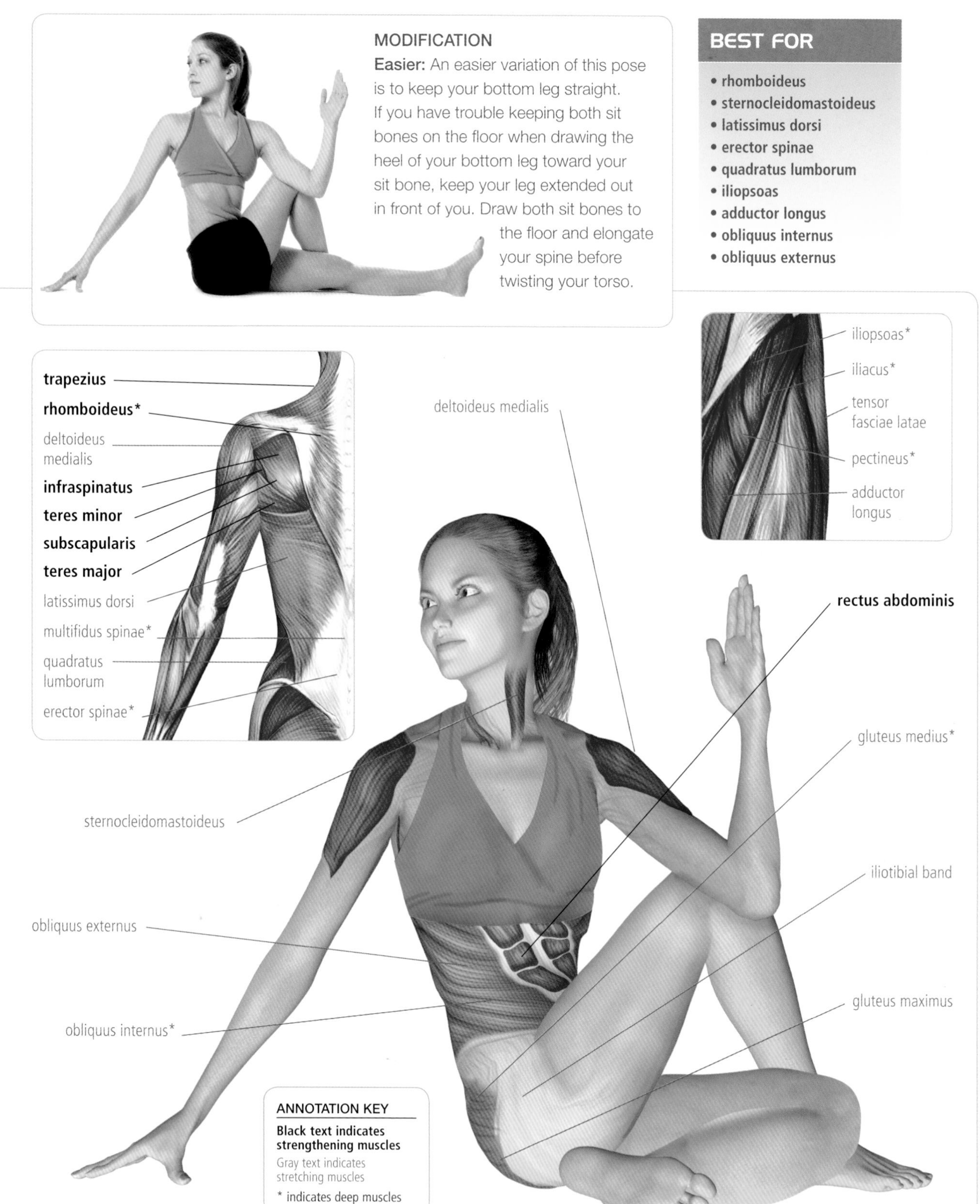

MODIFICATION

Easier: An easier variation of this pose is to keep your bottom leg straight. If you have trouble keeping both sit bones on the floor when drawing the heel of your bottom leg toward your sit bone, keep your leg extended out in front of you. Draw both sit bones to the floor and elongate your spine before twisting your torso.

BEST FOR

- **rhomboideus**
- **sternocleidomastoideus**
- **latissimus dorsi**
- **erector spinae**
- **quadratus lumborum**
- **iliopsoas**
- **adductor longus**
- **obliquus internus**
- **obliquus externus**

ANNOTATION KEY

Black text indicates strengthening muscles

Gray text indicates stretching muscles

* indicates deep muscles

TWISTING CHAIR POSE

(PARIVRTTA UTKATASANA)

1 Stand in Mountain Pose (Tadasana, see page 32), and then squat down in Awkward Pose (Utkatasana, see page 37), with your arms extended up toward the ceiling. Lean back slightly, so that your weight rests on your heels.

a

BEST FOR

- rectus abdominis
- obliquus internus
- transversus abdominis
- biceps femoris
- rectus femoris
- obliquus externus
- gluteus medius
- gluteus maximus

2 Squeezing your legs together, inhale, and bring your hands down to your chest. Press your palms together in prayer position.

3 Exhale, and twist toward the right, lengthening your spine as you remain in the squatting position. Rotate through your spine, torso, and shoulders, and place your left elbow on the outside of your right thigh. Look up toward the ceiling.

4 With each exhalation, deepen the twist, using your left elbow to guide your rotation.

5 Hold for 10 to 30 seconds. Inhale as you untwist, returning to Mountain Pose before twisting to the other side.

b

AVOID
- Lessening your squatted position as you twist.
- Forcing a deep twist too aggressively with your elbow.

PRONUNCIATION & MEANING
- Parivrtta Utkatasana (par-ee-vrt-tah OOT-kah-TAHS-anna)
- *parivrtta* = twist, revolve; *utkatasana* = chair

LEVEL
- Beginner

BENEFITS
- Stimulates digestion
- Stretches spine
- Strengthens thighs, buttocks, and abdominals

CONTRA-INDICATIONS & CAUTIONS
- Back injury

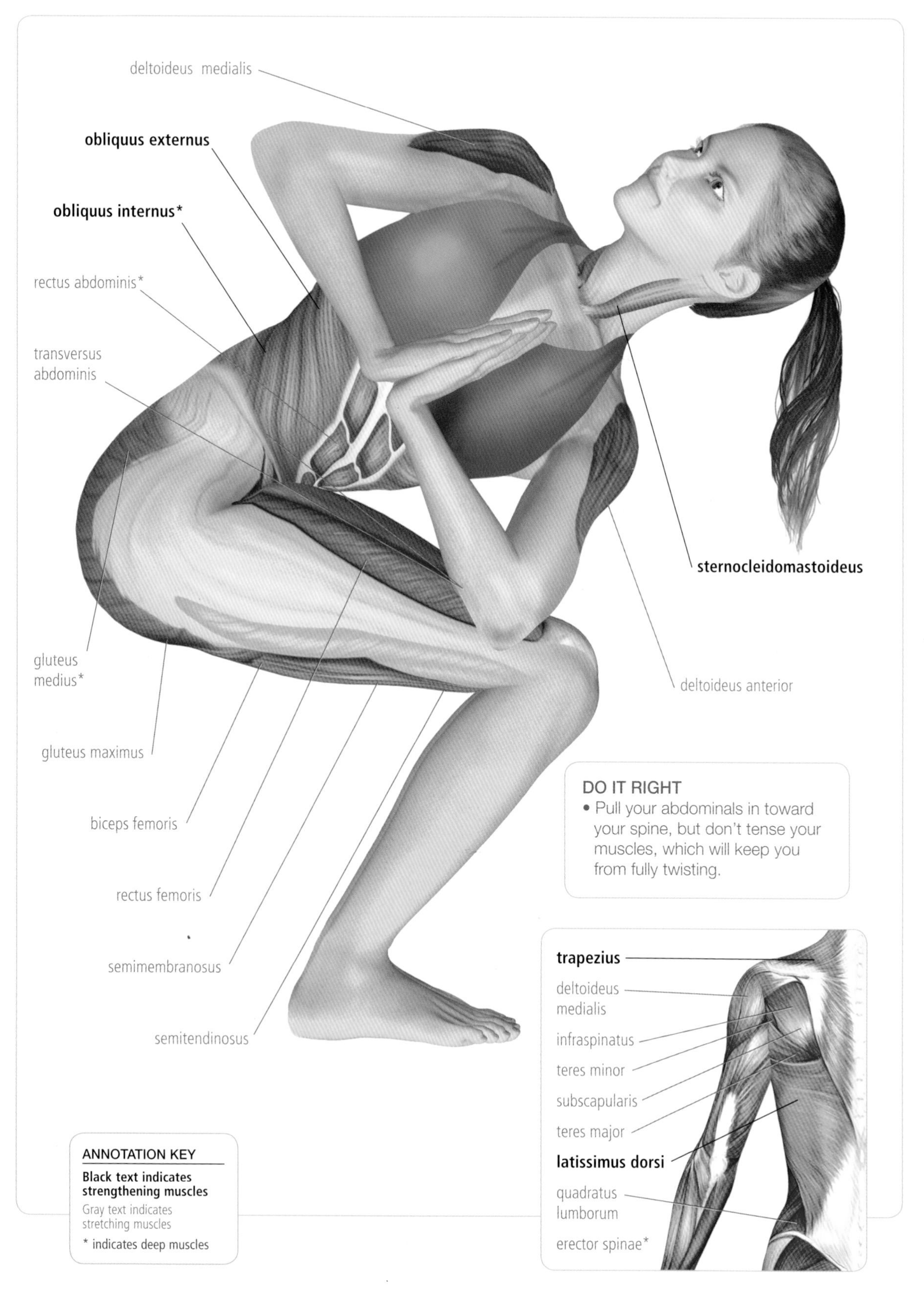

DO IT RIGHT
- Pull your abdominals in toward your spine, but don't tense your muscles, which will keep you from fully twisting.

ANNOTATION KEY
Black text indicates strengthening muscles
Gray text indicates stretching muscles
* indicates deep muscles

ARM SUPPORTS & INVERSIONS

As you age, your bones and upper-body strength deteriorate, increasing the risk of injury and making everyday tasks more difficult. Arm supports work to reverse the weakening of bones and muscles. These poses strengthen your arms, shoulders, and chest, and they help prevent osteoporosis. You will strengthen your abdominals as you engage them to balance and support your body. Arm supports do require a degree of flexibility, especially in your spine and hips. Let go of any unnecessary tension—the fear of falling on your face is natural. Overcome the fear by building your upper-body strength with diligent practice.

Inversions move your head below your heart, reversing the effects of gravity on your body. Inversions benefit the cardiovascular, lymphatic, nervous, and endocrine systems, increasing blood circulation and building healthier lung tissue. When beginning inversions, start by holding the poses for short periods, and be gentle on your neck.

UPWARD PLANK POSE

(PURVOTTANASANA)

a

1 Sitting in Staff Pose (Dandasana, see page 23) with your legs extended, place the palms of your hands on the floor several inches behind your hips, fingers facing forward.

DO IT RIGHT

- Use your hamstrings and shoulders to open your hips and chest, rather than overextend your back. If your hamstrings are too weak, keep your legs bent while holding the lift in your hips.
- Breathe steadily, using the breath to deepen the extension in your upper back.

b

2 Draw your knees toward your chest. Place your feet on the floor with your heels about 12 inches away from your buttocks, and turn your big toes slightly inward.

3 Exhale, pressing down with your hands and feet and lifting your hips until your back and thighs are parallel to the floor. Your shoulders should be directly above your wrists.

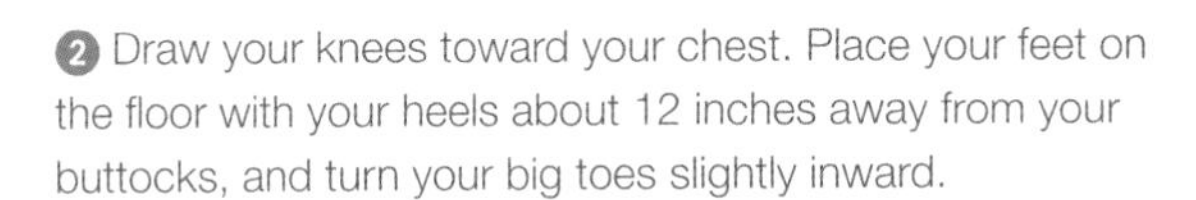

4 Without lowering your hips, straighten your legs one at a time.

5 Lifting your chest and bringing your shoulder blades together, push your hips higher, creating a slight arch in your back. Do not squeeze your buttocks to create the lift.

6 Slowly and gently elongate your neck and let it drop back.

7 Hold for 30 seconds and return to Staff Pose.

c

PRONUNCIATION & MEANING

- Purvottanasana (POOR-vo-tan-AHS-ahna)
- *purva* = front, east; *ut* = intense; *tan* = extend, stretch

LEVEL

- Intermediate

BENEFITS

- Strengthens the spine, arms, and hamstrings
- Extends the hips and chest

CONTRA-INDICATIONS & CAUTIONS

- Neck injury
- Wrist injury

BEST FOR

- deltoideus
- triceps brachii
- teres major
- teres minor
- erector spinae
- gluteus maximus
- gluteus medius
- adductor magnus
- biceps femoris

AVOID
- Using your buttocks muscles to maintain the position.
- Sagging your hips.

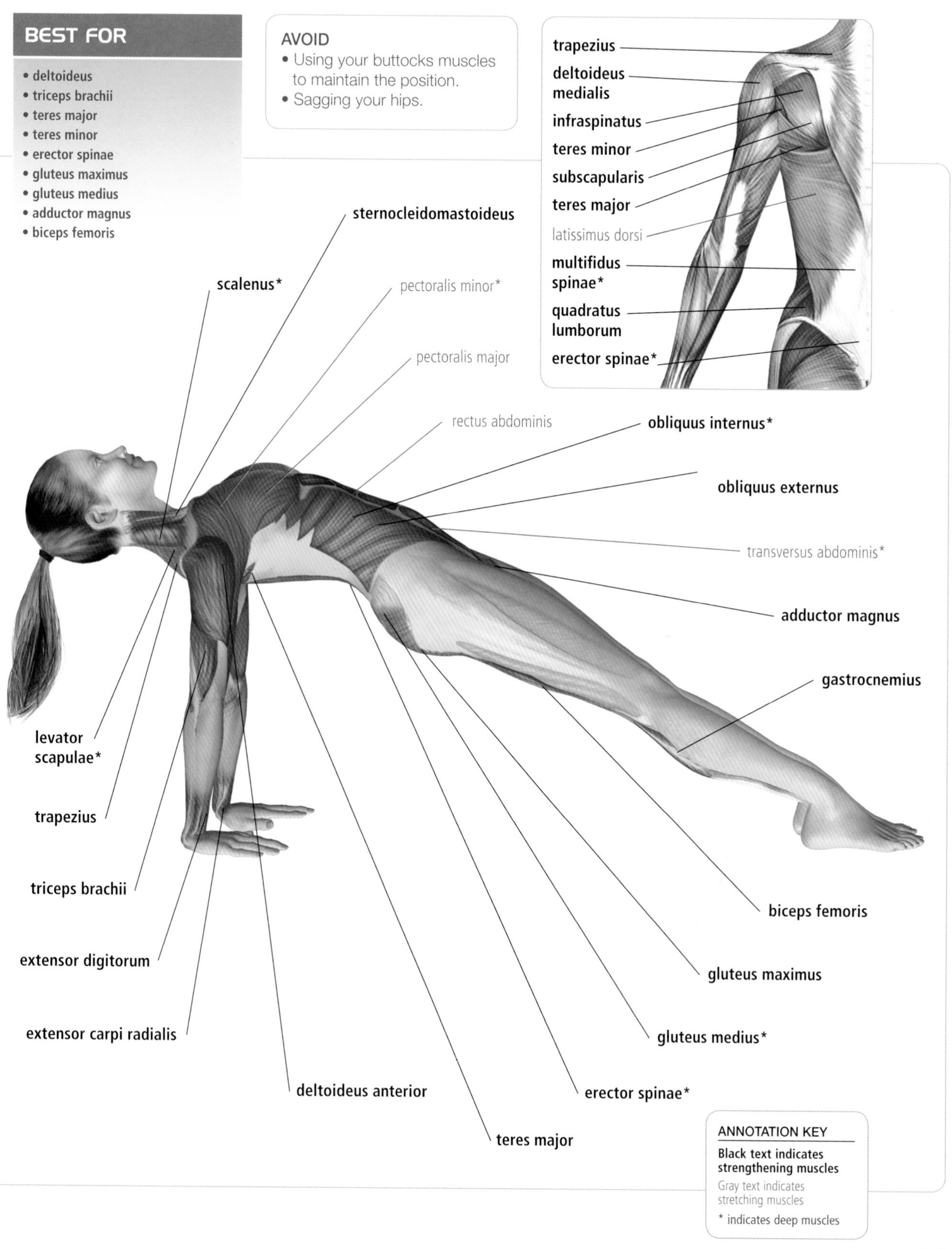

ANNOTATION KEY
Black text indicates strengthening muscles
Gray text indicates stretching muscles
* indicates deep muscles

CRANE POSE
(BAKASANA)

a

1 Begin in Garland Pose (Malasana, see pages 34–35), squatting with your feet and knees separated wider than your hips.

2 Lean your torso forward, and extend your arms to place your hands on the floor in front of you. Turn your hands inward slightly, and widen your fingers.

3 Bend your elbows, resting your knees against your upper arms. Lifting up on the balls of your feet and leaning forward with your torso, bring your thighs toward your chest and your shins to your upper arms. Round your back as you feel your weight transfer to your wrists.

b

4 Exhale, and slowly lift your feet off the floor one at a time. Keep your head position neutral, and find your balance point.

5 Hold for 20 seconds to 1 minute.

c

PRONUNCIATION & MEANING
- Bakasana (bak-AHS-anna)
- *baka* = crane, heron
- Also called Crow Pose

LEVEL
- Intermediate

BENEFITS
- Strengthens and tones arms and abdominals
- Strengthens wrists
- Improves balance

CONTRA-INDICATIONS & CAUTIONS
- Carpal tunnel syndrome
- Pregnancy

AVOID
- Dropping your head.
- Jumping into the posture.

DO IT RIGHT
- If you are afraid of falling forward, place a blanket in front of you as a cushion.
- Gaze at a spot on the floor in front of you to maintain balance.

serratus anterior
obliquus externus
obliquus internus*
rectus abdominis
transversus abdominis*

BEST FOR
- iliopsoas
- iliacus
- trapezius
- serratus anterior
- deltoideus
- triceps brachii
- biceps brachii
- coracobrachialis
- pectoralis major

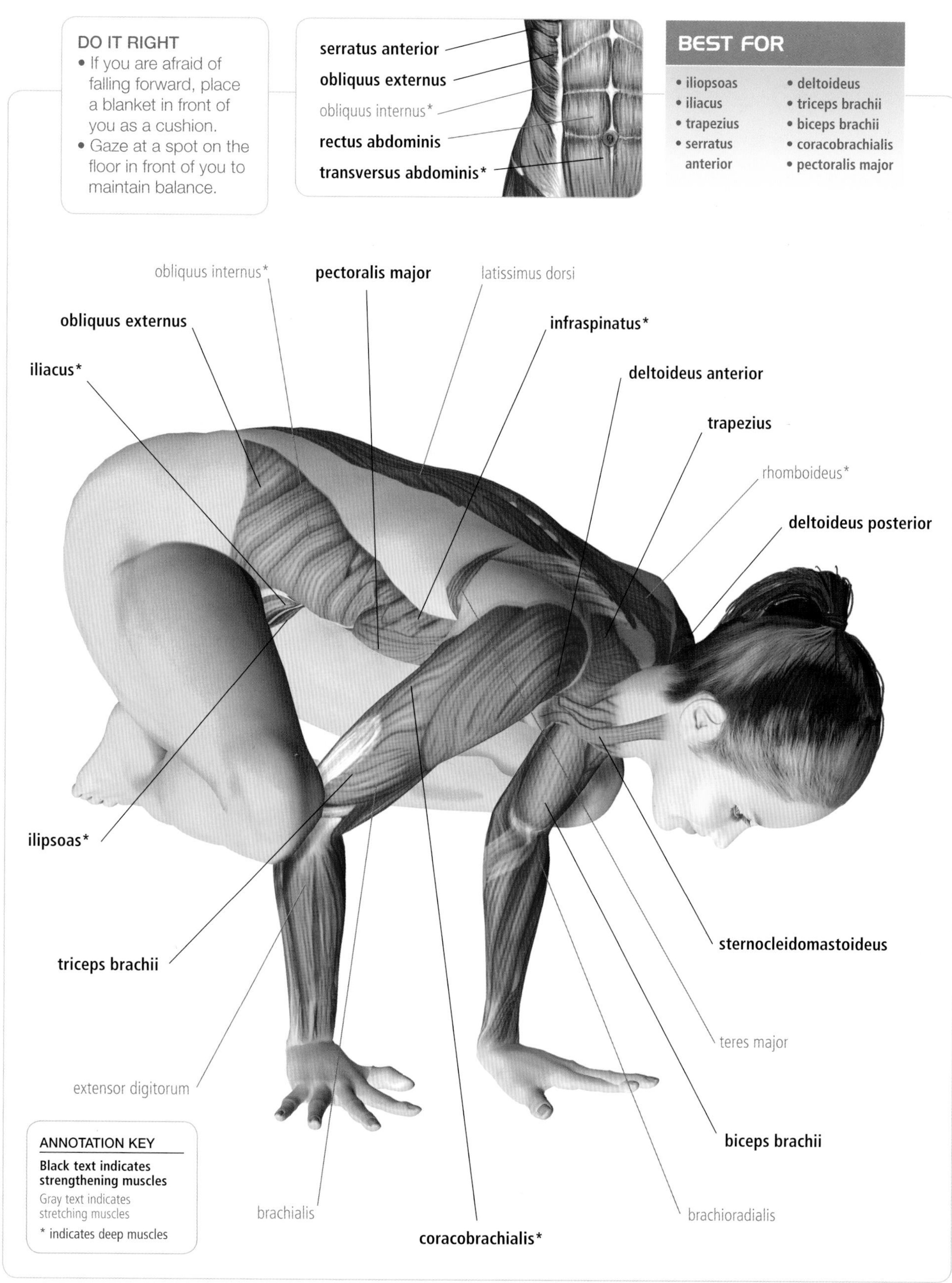

ANNOTATION KEY
Black text indicates strengthening muscles
Gray text indicates stretching muscles
* indicates deep muscles

SIDE CRANE POSE

(PARSVA BAKASANA)

1. Stand in Prayer Pose (Samasthiti, see page 33), with your hands together at the middle of your chest. With your legs together, begin by squatting deeply until your buttocks are just above your heels, which are lifted off the floor.

2. Bring your arms across your body to your right side, touching your left elbow to your right thigh as your hands reach the floor. Exhale, and deepen the twist, pulling your right shoulder back.

AVOID

- Dropping your head.
- Jumping into the posture.

3. Place your left hand flat on the floor outside your right thigh. Place the outside of your right thigh on your left upper arm. Lean to the right until you can place your right hand flat on the floor so that your hands are shoulder width apart. Your hips and shoulders should maintain a deep twist.

4. Slowly lift your pelvis as you shift your weight toward your hands, using your left arm as a support for your right thigh. Continue shifting to the right, drawing your abdominals in toward your spine. Keep your feet together as you raise them completely off the floor toward your buttocks, exhaling.

5. Hold for 20 seconds to 1 minute, breathing through the balance. Exhale as you bring your feet to the floor. Repeat on the other side.

PRONUNCIATION & MEANING

- Parsva Bakasana (parsh-vah bak-AHS-anna)
- *baka* = crane, heron; *parsva* = side
- Also called Side Crow Pose

LEVEL

- Advanced

BENEFITS

- Strengthens and tones arms and abdominals
- Strengthens wrists
- Improves balance

CONTRA-INDICATIONS & CAUTIONS

- Wrist injury
- Lower-back injury

BEST FOR

- iliopsoas
- iliacus
- trapezius
- serratus anterior
- deltoideus
- triceps brachii
- biceps brachii
- coracobrachialis
- pectoralis major
- obliquus internus

DO IT RIGHT
- If you are afraid of falling forward, place a blanket in front of you as a cushion.
- Gaze at a spot on the floor in front of you to maintain balance.
- Focus on twisting deeply as you go into the posture.

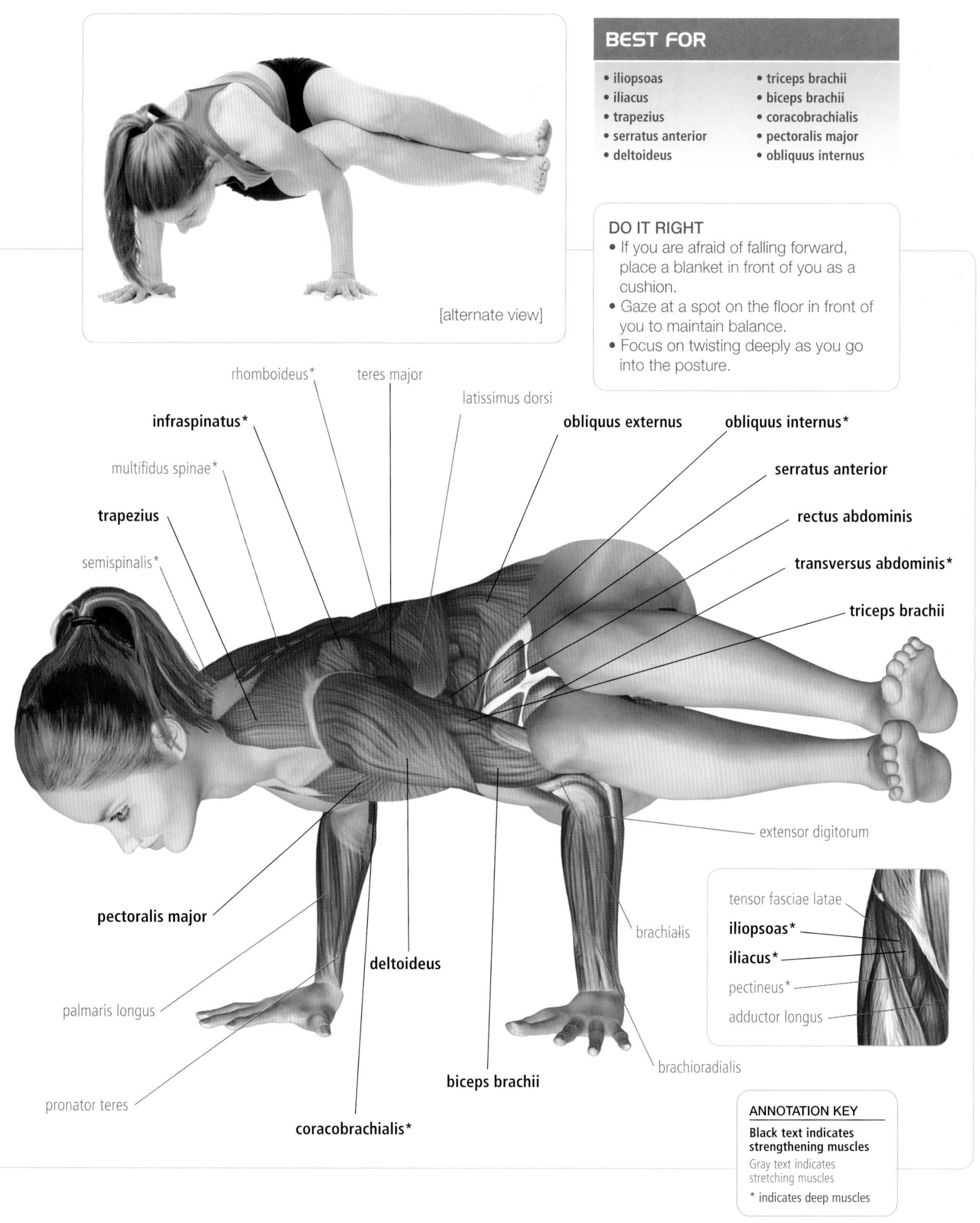

ANNOTATION KEY
Black text indicates strengthening muscles
Gray text indicates stretching muscles
* indicates deep muscles

PLANK POSE TO

❶ To assume Plank Pose, begin in Downward-Facing Dog (Adho Mukha Svanasana, see page 24).

❷ Inhale, and draw your torso forward until your wrists are directly under your shoulders at a 90-degree angle. Your body should form a straight line from the top of your head to your heels.

❸ Press your hands firmly down into the floor, and, not letting your chest sink, press back through your heels.

a

DO IT RIGHT
- Lengthen your legs all the way through your heels to evenly distribute weight while in Plank Pose.
- Squeeze your buttocks muscles, and draw in your abdominals for stability.

AVOID
- Sinking your shoulders.
- Sagging your hips or raising your buttocks.
- Hunching your shoulders up toward your ears.

❹ Keeping your neck in line with your spine, broaden your shoulder blades. Your legs should be strong, straight, and engaged, and you're your feet should be square, with your heels pointing upward toward the ceiling. Hold for 30 seconds to 1 minute.

b

❺ From Plank Pose, open your chest, and broaden your shoulder blades while tucking in your tailbone.

❻ Exhale, and with your legs turned slightly inward, lower yourself to the floor until your upper arms are parallel to your spine.

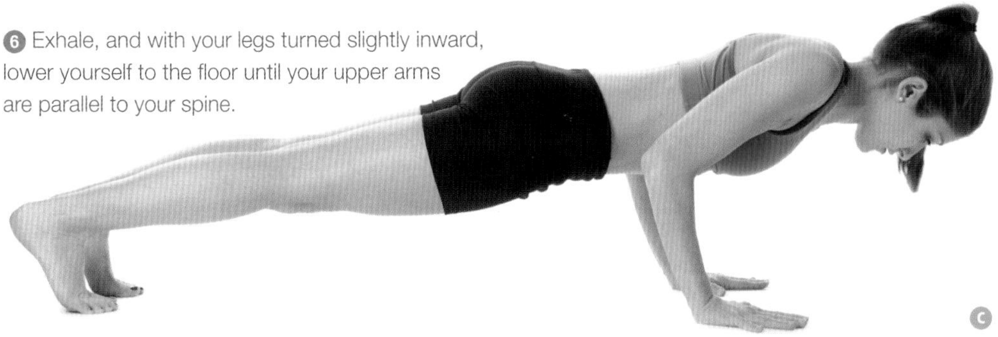

c

PRONUNCIATION & MEANING
- There is no agreed-upon Sanskrit name for the Plank Pose.
- Chaturanga Dandasana (chaht-tour-ANG-ah don-DAHS-anna)
- *chatur* = four; *anga* = limb; *danda* = staff, stick

LEVEL
- Beginner/ Intermediate

BENEFITS
- Strengthens and tones arms and abdominals
- Strengthens wrists

CONTRA-INDICATIONS & CAUTIONS
- Shoulder issues
- Wrist injury
- Lower-back injury

FOUR-LIMBED STAFF POSE

(CHATURANGA DANDASANA)

❼ Tuck your tailbone under, and draw your abdominals in toward your spine to maintain the straight line from your shoulders to your heels. Keep your elbows in by your sides. Lift your head and look forward.

❽ Hold for 10 to 30 seconds.

BEST FOR

- rectus abdominis
- triceps brachii
- subscapularis
- supraspinatus
- infraspinatus
- teres major
- pectoralis major
- pectoralis minor

DO IT RIGHT

- If you find it too difficult to support yourself in Four-limbed Staff Pose, begin with the Plank Pose, and then place your knees on the floor. Continue kneeling, exhale, and lower your torso toward the floor until there is just an inch or two between your chest and the floor.

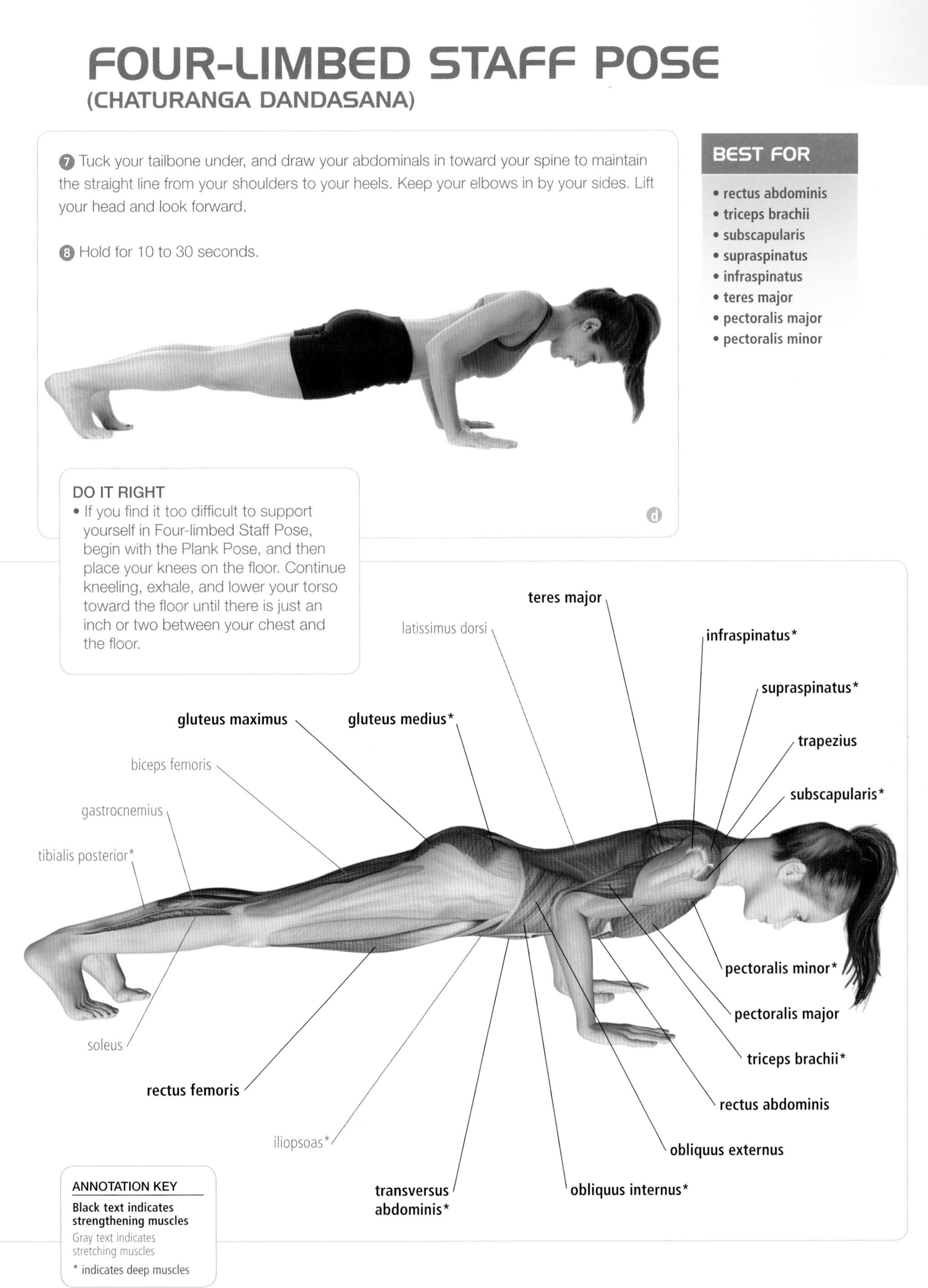

ANNOTATION KEY

Black text indicates strengthening muscles

Gray text indicates stretching muscles

* indicates deep muscles

EIGHT-ANGLE POSE
(ASTAVAKRASANA)

(a)

1. Sitting on the floor, open your hips, allowing your knees to lower toward the floor.

2. Lift your right leg so that it is bent, with your thigh perpendicular to the floor. Use your arms to pull your right leg over your right shoulder. The back of your knee should rest on top of your shoulder.

3. Lean your torso forward, and place your hands shoulder-width apart on the floor in front of you. Your right hand should be placed on the outside of the right leg.

4. Shift your weight forward onto your hands, and press up, lifting your chest through the movement. Straighten your left leg in front of you.

(b)

5. Exhale, and lower your torso until it is parallel to the floor. Draw your left leg toward the right. Bending both legs so that they lock at the ankles, hook your right ankle below the left.

(c)

6. Bend your arms and lower your chest toward the floor, squeezing your legs together and extending them to the right. Your thighs should squeeze your right arm and should be parallel to the floor.

7. Twist your torso to the left, and keep your elbows in by your sides. Look forward, gazing at the floor in front of you.

8. Hold for 30 seconds to 1 minute. Slowly straighten your arms and lift your torso. Bend your knees, unhook your ankles, and return to a seated position on the floor. Repeat on the other side.

(d)

PRONUNCIATION & MEANING
- Astavakrasana (ah-SHTA-vak-RAHS-anna)
- *ashta* = eight; *vakra* = bent, curved

LEVEL
- Advanced

BENEFITS
- Strengthens wrists, arms, and abdominals
- Increases balance and flexibility

CONTRA-INDICATIONS & CAUTIONS
- Shoulder issues
- Wrist injury
- Elbow injury

AVOID
- Allowing your top hip to rock backward, causing your bottom hip to drop.

[alternate view]

BEST FOR

- adductor magnus
- adductor longus
- triceps brachii
- biceps brachii

DO IT RIGHT
- To keep your legs symmetrical, twist more from the spine than from your hips.
- If you struggle with keeping your body lifted off the floor, use blocks for your hands to practice pressing your hips up as one leg rests on your shoulder.

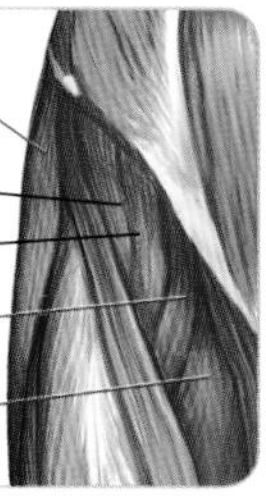

tensor fasciae latae
iliopsoas*
iliacus*
pectineus*
adductor longus

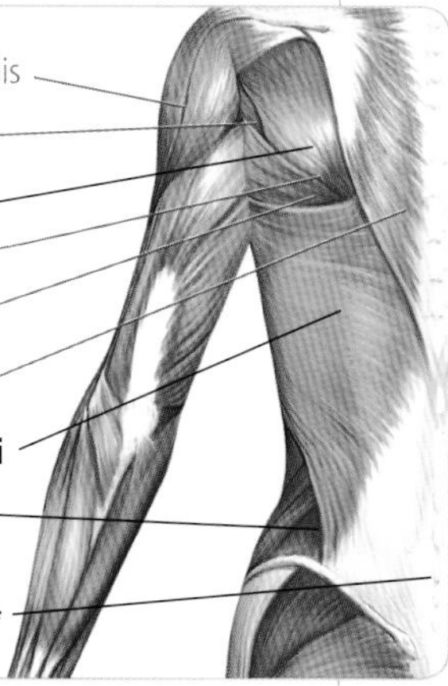

sternocleidomastoideus
trapezius
deltoideus anterior
serratus anterior
gastrocnemius
semimembranosus
triceps brachii
transversus abdominis*
vastus intermedius*
adductor magnus
adductor longus
semitendinosus
soleus
tibialis anterior
biceps brachii
rectus abdominis
pectoralis major
scalenus*

deltoideus medialis
teres minor
infraspinatus*
subscapularis
teres major
rhomboideus
latissimus dorsi
quadratus lumborum
erector spinae*

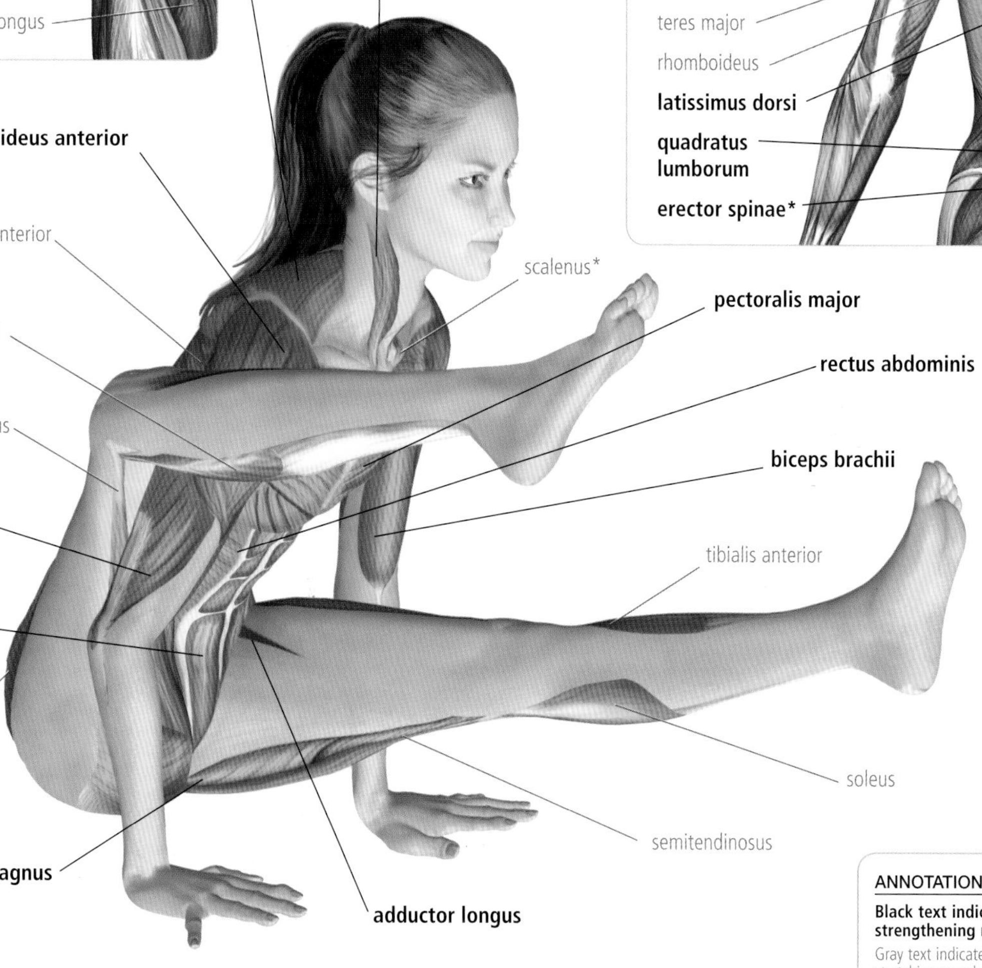

ANNOTATION KEY
Black text indicates strengthening muscles
Gray text indicates stretching muscles
* indicates deep muscles

SIDE PLANK POSE
(VASISTHASANA)

1 Begin in Plank Pose (see page 134). Your arms should be straight with your wrists aligned under the shoulder. To prepare for the Side Plank Pose, you may want your hands slightly in front of your shoulders to push into the support.

a

2 Shift your weight onto the outside of your right foot and onto your right arm. Roll to the side, guiding with your hips and bringing your left shoulder back. Stack your left foot on top of the right, squeezing both legs together and straight.

b

AVOID
- Allowing your hips or shoulders to sway or sink.
- Lifting your hips too high.

3 Exhale, bring the left arm up to the ceiling, and elongate your body, making a straight line from your head to your heels. Gaze up at your fingertips as you continue to push through your shoulder into the floor, maintaining a strong balance.

c

4 Breathe, and hold the posture for 15 to 30 seconds. Return to Plank Pose or Downward-Facing Dog (Adho Mukha Svanasana, see page 24), and repeat on the left side.

d

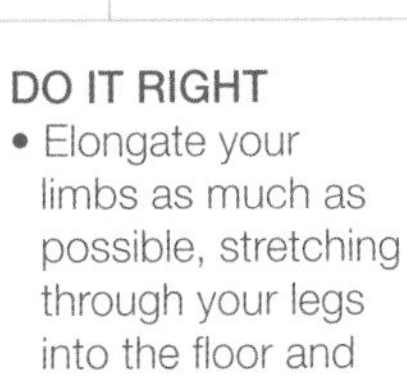

PRONUNCIATION & MEANING
- Vasisthasana (vah-sish-TAHS-anna)
- *vasistha* = most excellent, best, richest

LEVEL
- Beginner

BENEFITS
- Strengthens wrists, arms, legs, and abdominals
- Improves balance

CONTRA-INDICATIONS & CAUTIONS
- Shoulder issues
- Wrist injury
- Elbow injury

DO IT RIGHT
- Elongate your limbs as much as possible, stretching through your legs into the floor and reaching your top arm high to the ceiling.
- Your feet should be stacked and flexed as if they were side by side in standing position.

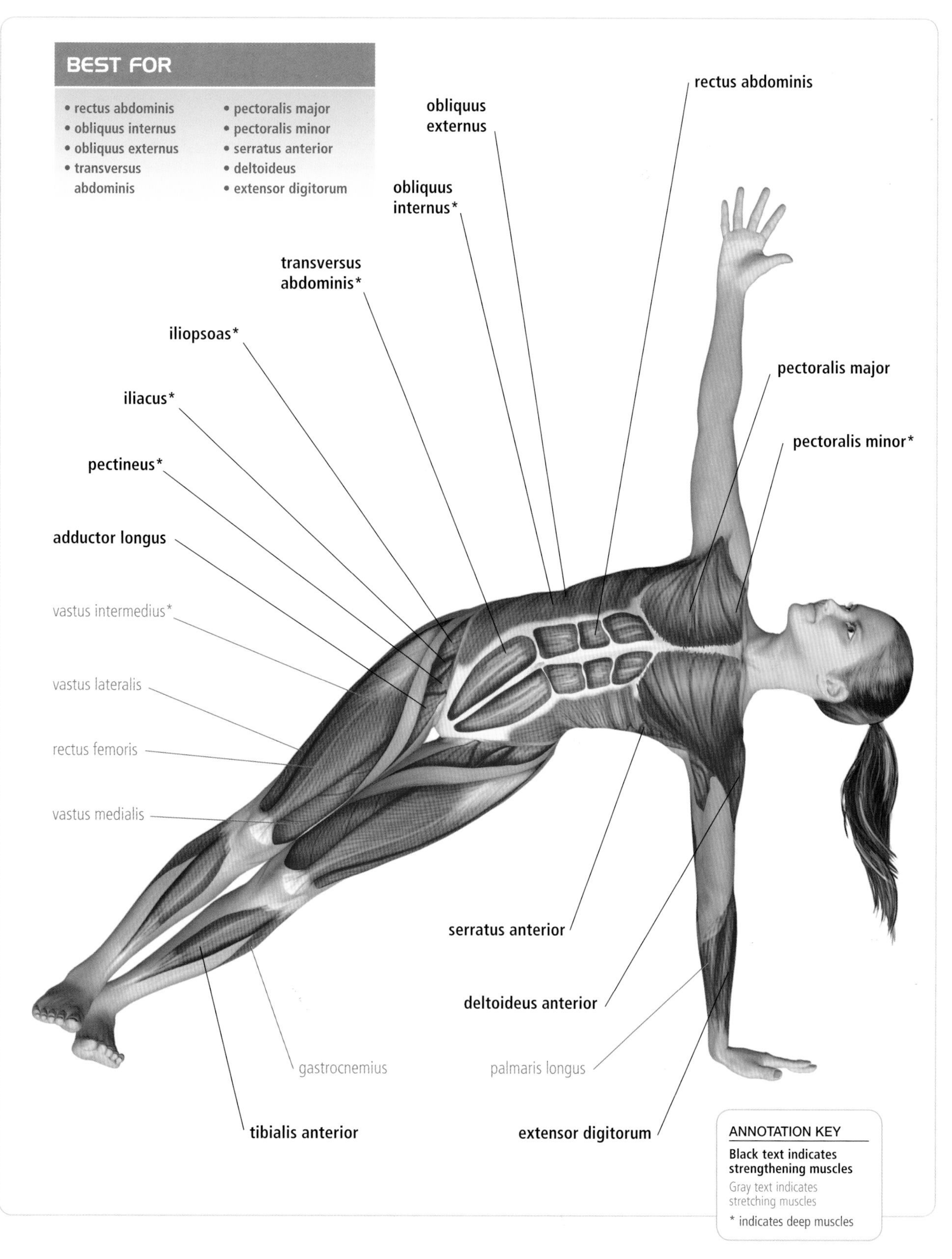
BEST FOR
• rectus abdominis
• obliquus internus
• obliquus externus
• transversus abdominis
• pectoralis major
• pectoralis minor
• serratus anterior
• deltoideus
• extensor digitorum
rectus abdominis
obliquus externus
obliquus internus*
transversus abdominis*
iliopsoas*
iliacus*
pectineus*
adductor longus
vastus intermedius*
vastus lateralis
rectus femoris
vastus medialis
pectoralis major
pectoralis minor*
serratus anterior
deltoideus anterior
palmaris longus
gastrocnemius
tibialis anterior
extensor digitorum
ANNOTATION KEY
Black text indicates strengthening muscles
Gray text indicates stretching muscles
* indicates deep muscles

PLOW POSE

(HALASANA)

1. Lie supine on the floor with your knees bent. Your arms should be by your sides, with your hands placed flat on the floor.

2. Tighten your abdominals, and lift your knees off the floor. Exhale, press your arms into the floor, and lift your knees higher so that your buttocks and hips come off the floor.

3. Continue lifting your knees toward your face, and roll your back off the mat from your hips to your shoulders. With your upper arms firmly planted on the floor, bend your elbows, and place your hands on your lower back. Draw your elbows in closer to your sides.

4. Inhale, tuck your tailbone toward your pubis, and straighten your legs back toward your head. Your torso should be perpendicular to the floor.

5. Exhale, and continue to extend you legs beyond your head. Squeeze your legs and bend at your waist until your toes touch the floor. Place your hands face down on the floor, pushing through your arms to maintain the lift in your hips.

6. Hold for 1 to 5 minutes.

AVOID
- Swinging your legs down quickly into the pose.

PRONUNCIATION & MEANING
- Halasana (hah-LAHS-anna)
- *hala* = plow

LEVEL
- Intermediate

BENEFITS
- Relieves stress
- Relieves backache and headache
- Stimulates digestion

CONTRA-INDICATIONS & CAUTIONS
- High blood pressure
- Neck issues
- Menstruation or pregnancy

DO IT RIGHT
- Soften your throat, and relax your tongue.
- If you are uncomfortable moving your hands from your back or your toes don't reach the floor, continue supporting your back with your hands.
- Place folded blankets below your shoulders if the posture strains your neck.

BEST FOR

- rectus abdominis
- latissimus dorsi
- transversus abdominis
- triceps brachii
- infraspinatus
- supraspinatus
- subscapularis

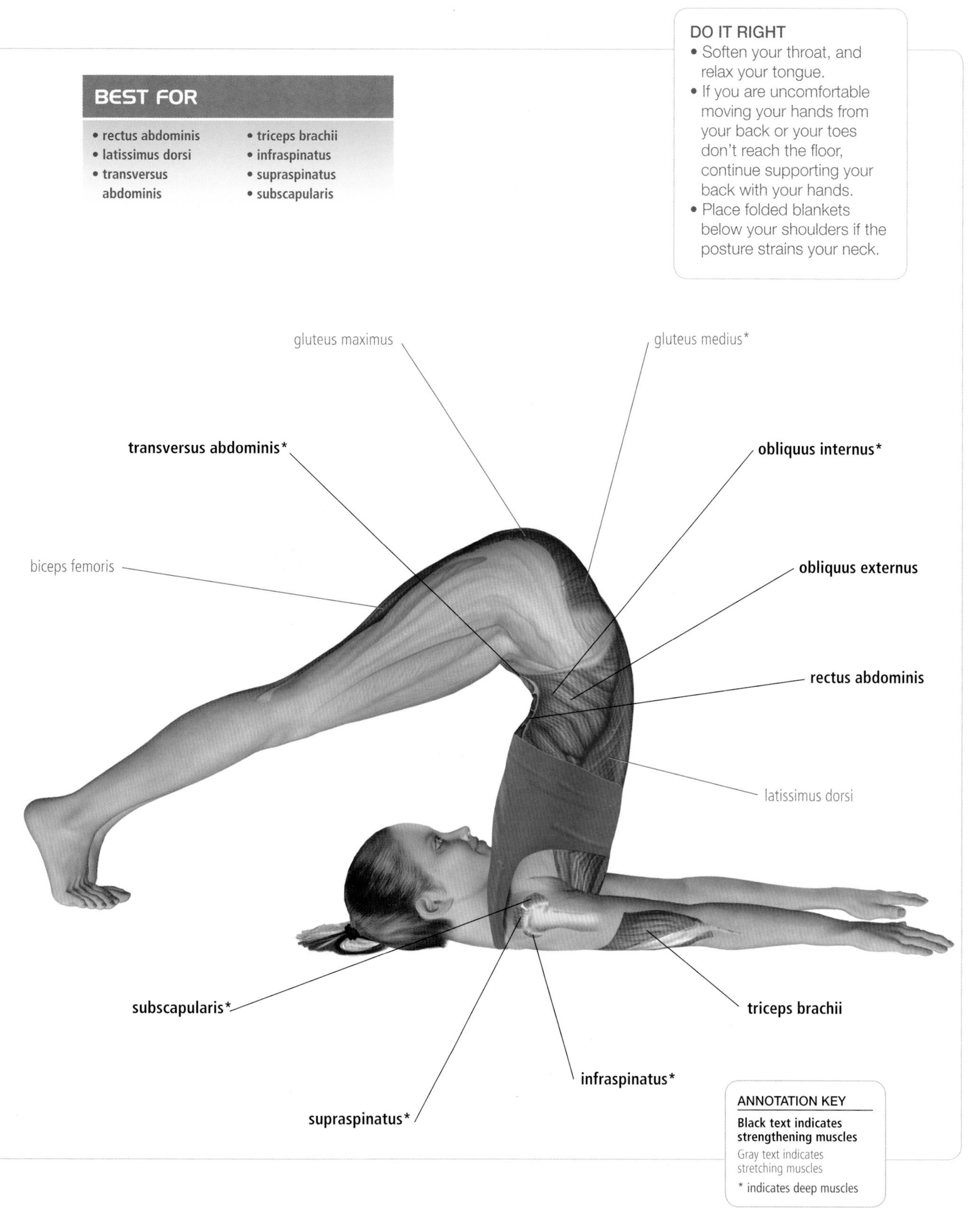

ANNOTATION KEY
Black text indicates strengthening muscles
Gray text indicates stretching muscles
* indicates deep muscles

SUPPORTED SHOULDERSTAND

(SALAMBA SARVANGASANA)

1. Lie supine on the floor with your knees bent and arms by your sides.

2. Tighten your abdominals, and lift your knees off the floor. Exhale, press your arms into the floor, and lift your knees higher so that your buttocks come off the floor.

3. Continue lifting your knees toward your face, and roll your back off the mat from your hips to your shoulders. With your upper arms firmly planted on the floor, bend your elbows, and place your hands on your lower back. Draw your elbows in closer to your sides.

4. Inhale, tuck your tailbone toward your pubis, and straighten your legs back toward your head. Your torso should be perpendicular to the floor.

5. With your next inhalation, extend your legs up toward the ceiling, opening your hips as you lift. Squeeze your buttocks, and press down with your elbows to create a straight, elongated line from your chest to your toes.

6. Hold for 30 seconds to 5 minutes before bending your knees and hips and returning to the floor.

AVOID
- Bending at your hips once you are in the posture, because it puts added pressure on your neck and spine.
- Splaying your elbows out to the sides.

PRONUNCIATION & MEANING
- Salamba Sarvangasana (sah-LOM-bah sar-van-GAHS-anna)
- *sa* = with; alamba = support; *sarva* = all; *anga* = limb

LEVEL
- Intermediate

BENEFITS
- Relieves stress
- Stretches shoulders, neck, and upper spine
- Stimulates digestion

CONTRA-INDICATIONS & CAUTIONS
- High blood pressure
- Neck issues
- Headache or ear infection

DO IT RIGHT
- Soften your throat, and relax your tongue.
- If you can't lift your pelvis into the inversion, practice a few feet away from a wall and walk your feet up the wall until you can place your hands on your back.
- Place folded blankets below your shoulders if the posture strains your neck.

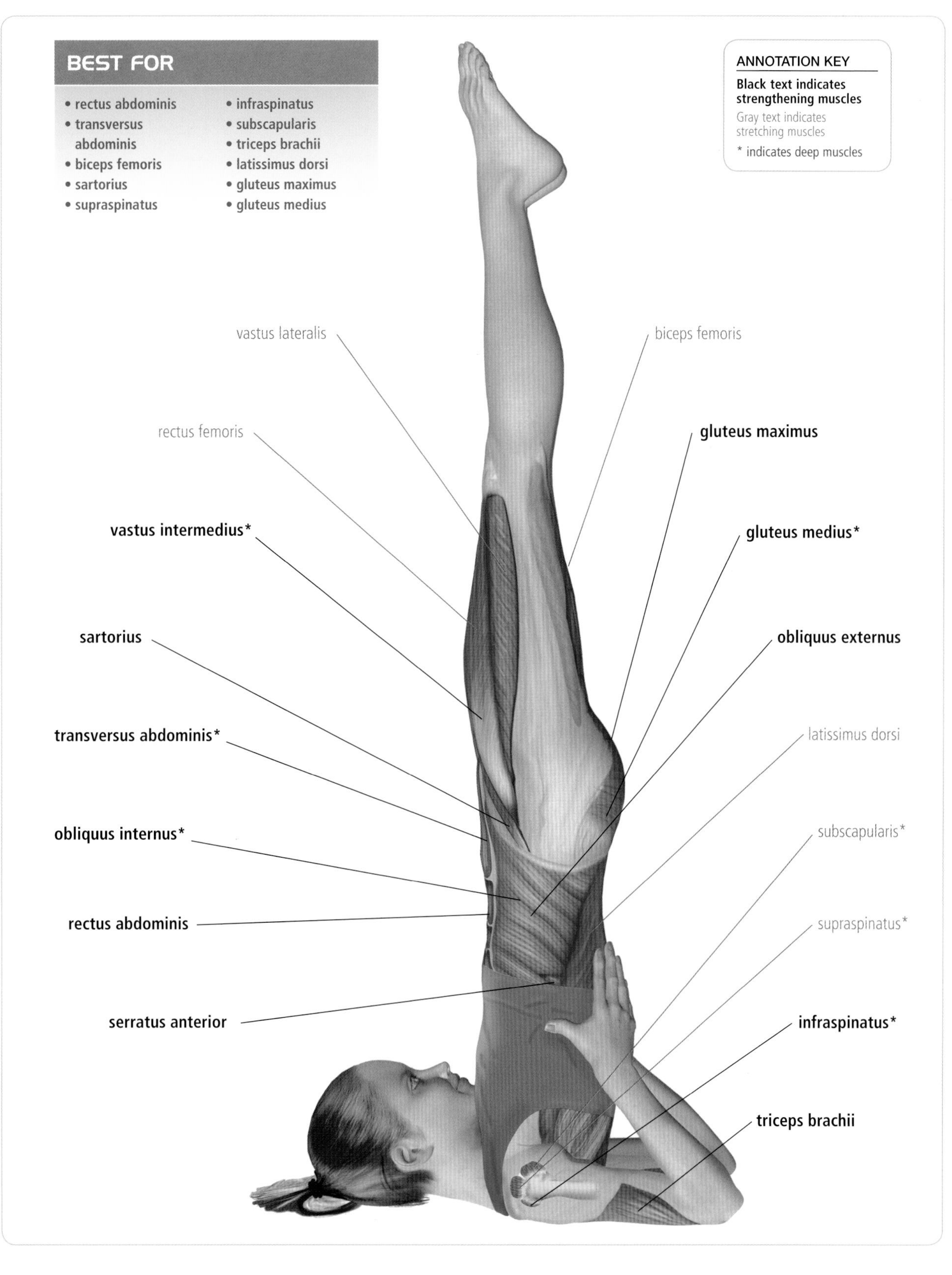
BEST FOR
• rectus abdominis
• transversus abdominis
• biceps femoris
• sartorius
• supraspinatus
• infraspinatus
• subscapularis
• triceps brachii
• latissimus dorsi
• gluteus maximus
• gluteus medius
ANNOTATION KEY
Black text indicates strengthening muscles
Gray text indicates stretching muscles
* indicates deep muscles
vastus lateralis
biceps femoris
rectus femoris
gluteus maximus
vastus intermedius*
gluteus medius*
sartorius
obliquus externus
transversus abdominis*
latissimus dorsi
obliquus internus*
subscapularis*
rectus abdominis
supraspinatus*
serratus anterior
infraspinatus*
triceps brachii

DOLPHIN POSE TO

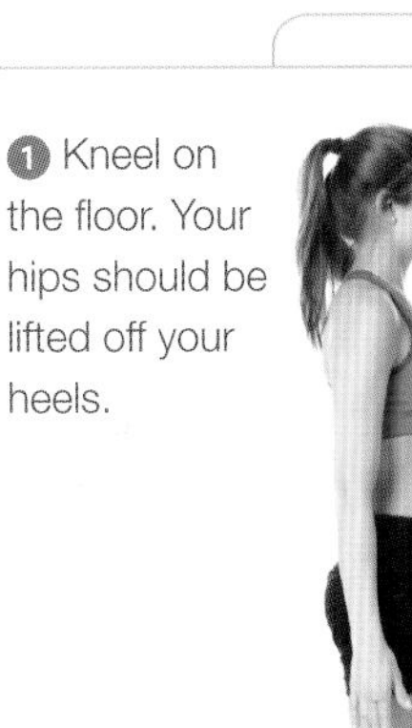

1 Kneel on the floor. Your hips should be lifted off your heels.

a

2 Place your hands on the floor in front of you and lower your elbows to the floor, keeping them in by your sides and aligned with your shoulders.

3 Exhale, and lift your knees off the floor. Ground your feet flat on the floor, pushing your heels down.

4 Straighten your legs as you lift your sit bones to the ceiling. Tuck your tailbone toward your pubis, and squeeze your legs together.

5 Push through your forearms, and extend the stretch through your shoulders. Keep your head and chest lifted off the floor.

6 Hold for 30 seconds to 1 minute.

b

7 From Dolphin Pose, inhale, and slowly walk toward your head on the balls of your feet, causing your hips to lift up toward the ceiling.

c

8 Once your weight has shifted to your shoulders and forearms, and your sit bones are pointed toward the ceiling, exhale, and lift your feet off the floor. Use your abdominals to slowly raise both toes simultaneously.

9 Lengthening your spine and keeping your shoulders open, bend your knees and draw your thighs toward your abdominals. Make sure that your torso remains perpendicular to the floor. Take a few breaths, and balance in this position.

d

PRONUNCIATION & MEANING
- There is no agreed-upon Sanskrit name for the Dolphin Pose.
- Salamba Sirsasana (sah-LOM-bah shear-SHAHS-anna)
- *sa* = with; *alamba* = support; *sirsa* = head

LEVEL
- Advanced

BENEFITS
- Strengthens and tones abdominals
- Strengthens arms, legs, and spine
- Improves balance

CONTRA-INDICATIONS & CAUTIONS
- Back injury
- Neck injury
- Headache
- High blood pressure

DO IT RIGHT
- While holding Dolphin Pose, keep your back straight. If you cannot straighten your legs without sagging or rounding your spine, keep your knees slightly bent.
- When in the headstand, support your weight evenly between your forearms.
- If you have trouble balancing in the headstand or distributing the majority of your weight onto your arms and shoulders, practice with the backs of your shoulders against a wall.

SUPPORTED HEADSTAND
(SALAMBA SIRSASANA)

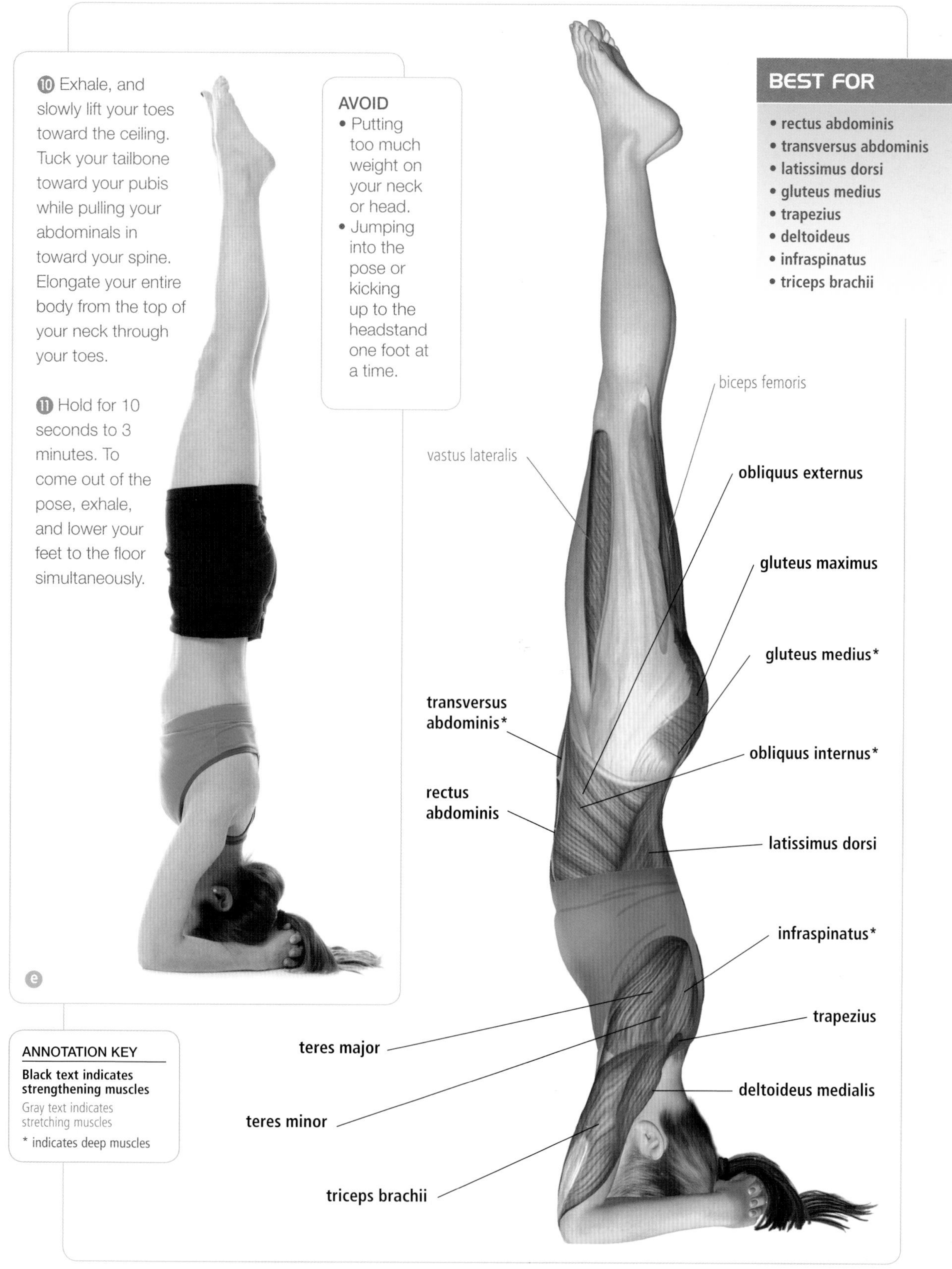

⑩ Exhale, and slowly lift your toes toward the ceiling. Tuck your tailbone toward your pubis while pulling your abdominals in toward your spine. Elongate your entire body from the top of your neck through your toes.

⑪ Hold for 10 seconds to 3 minutes. To come out of the pose, exhale, and lower your feet to the floor simultaneously.

AVOID
- Putting too much weight on your neck or head.
- Jumping into the pose or kicking up to the headstand one foot at a time.

BEST FOR
- rectus abdominis
- transversus abdominis
- latissimus dorsi
- gluteus medius
- trapezius
- deltoideus
- infraspinatus
- triceps brachii

ANNOTATION KEY

Black text indicates strengthening muscles

Gray text indicates stretching muscles

* indicates deep muscles

YOGA SEQUENCES

Familiarizing yourself with various yoga poses is only the first step in your yoga practice. Incorporating these poses into sequences, flowing from one pose to the next, allows you to maximize the strength and flexibility that you will gain throughout your entire body. Generally, yoga sequences begin with gentler poses, build up to those that are more challenging, and end with a cool down. It's best to start your daily routine with several rounds of Sun Salutation. The following sequences in this chapter are a guide to get you started. With each individual pose, focus on attaining the proper body position before moving on. Combine other poses to add variety and create a yoga practice that best suits your body's needs.

SUN SALUTATION A

❶ Mountain Pose (Tadasana) page 32

❷ Upward Salute (Urdhva Hastasana) page 36

❸ Standing Forward Bend (Uttanasana) page 66

❹ Standing Half Forward Bend (Ardha Uttanasana) page 67

❺ Low Lunge Pose (Anjeneyasana) page 50

❻ Plank Pose page 134

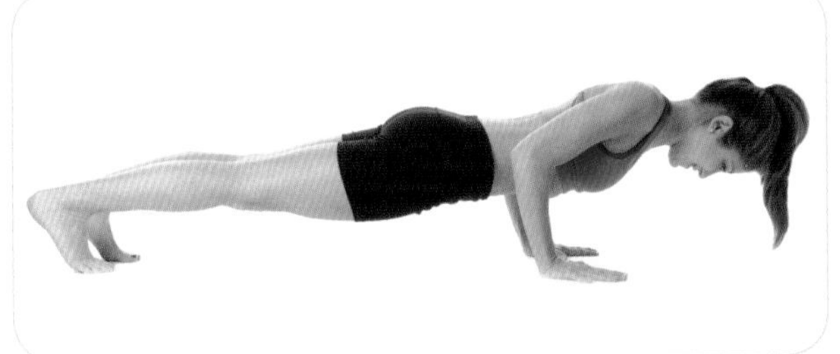

❼ Four-Limbed Staff Pose (Chaturanga Dandasana) page 135

❽ Upward-Facing Dog (Urdhva Mukha Svanasana) pages 78–79

❾ Downward-Facing Dog (Adho Mukha Svanasana) page 24

❿ Low Lunge, opposite leg (Anjeneyasana) page 50

⓫ Standing Half Forward Bend (Ardha Uttanasana) page 67

⓬ Standing Forward Bend (Uttanasana) page 66

⓭ Upward Salute (Urdhva Hastasana) page 36

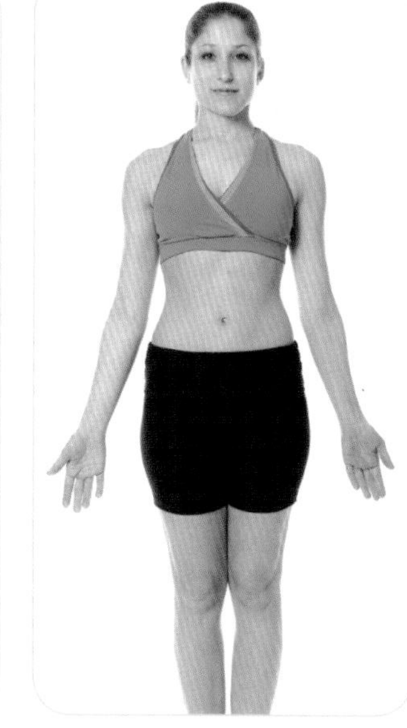

⓮ Mountain Pose (Tadasana) page 32

SUN SALUTATION B

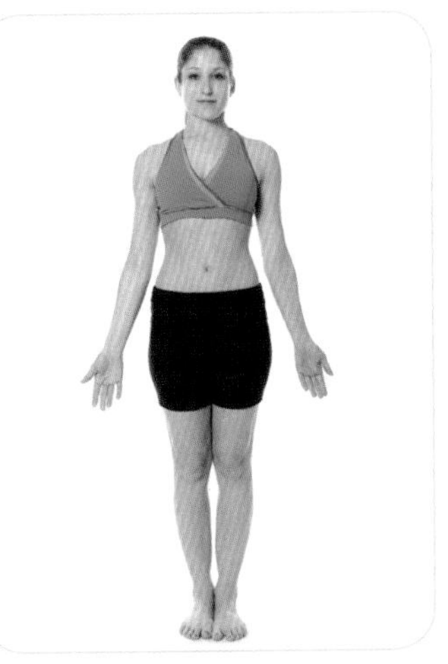

❶ Mountain Pose (Tadasana) page 32

❷ Awkward Pose (Utkatasana) page 37

❸ Standing Forward Bend (Uttanasana) page 66

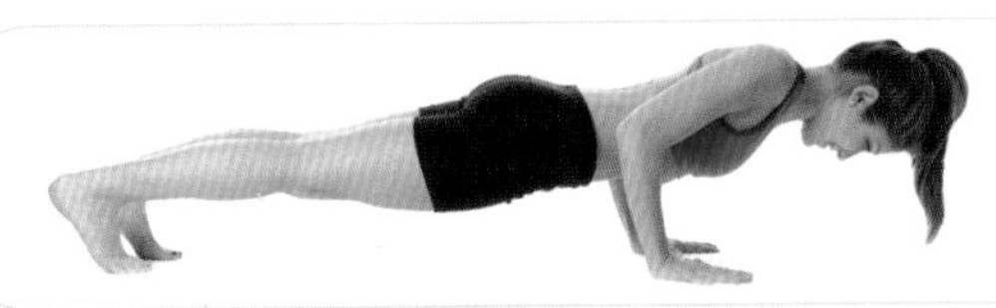

❹ Four-Limbed Staff Pose (Chaturanga Dandasana) page 135

❺ Upward-Facing Dog (Urdhva Mukha Svanasana) pages 78–79

❻ Downward-Facing Dog (Adho Mukha Svanasana) page 24

❼ Warrior I (Virabhadrasana I) pages 54–55

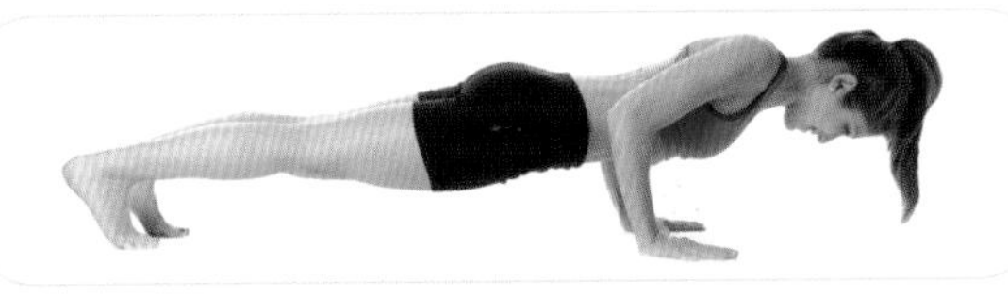

❽ Four-Limbed Staff Pose (Chaturanga Dandasana) page 135

❾ Upward-Facing Dog (Urdhva Mukha Svanasana) pages 78–79

⑩ Downward-Facing Dog (Adho Mukha Svanasana) page 24

⑪ Warrior I, opposite leg (Virabhadrasana I) pages 54–55

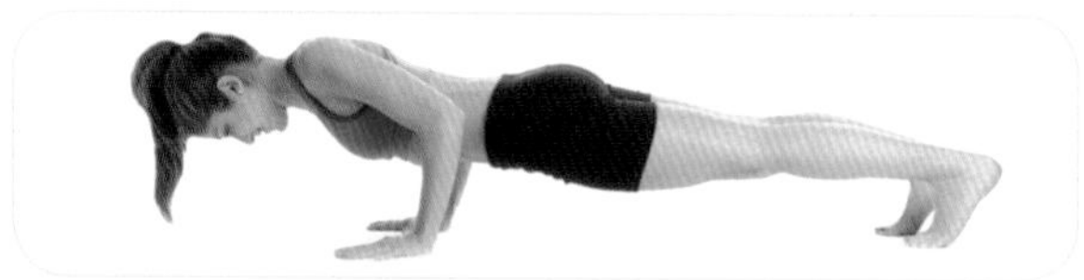

⑫ Four-Limbed Staff Pose (Chaturanga Dandasana) page 135

⑬ Upward-Facing Dog (Urdhva Mukha Svanasana) pages 78–79

⑭ Downward-Facing Dog (Adho Mukha Svanasana) page 24

⑮ Standing Forward Bend (Uttanasana) page 66

⑯ Awkward Pose (Utkatasana) page 37

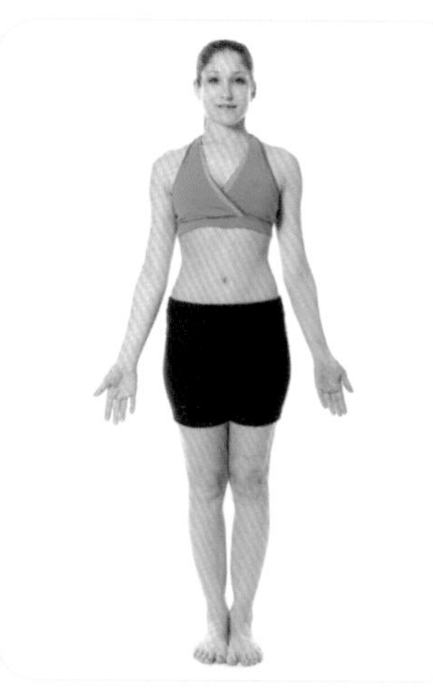

⑰ Mountain Pose (Tadasana) page 32

BEGINNER SEQUENCE

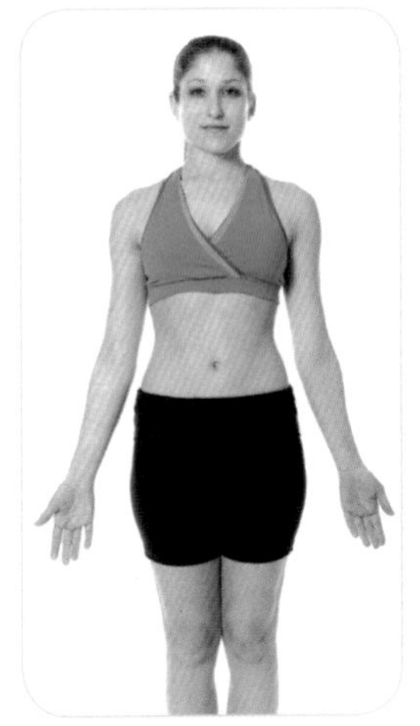

❶ Mountain Pose (Tadasana) page 32

❷ High Lunge (pages 52–53)

❸ Downward-Facing Dog (Adho Mukha Svanasana) page 24

❹ Warrior I (Virabhadrasana I) pages 54–55

❺ Intense Side Stretch (Parsvottanasana) pages 64–65

❻ Tree Pose (Vrksasana) pages 38–39

❼ Awkward Pose (Utkatasana) page 37

❽ Downward-Facing Dog (Adho Mukha Svanasana) page 24

❾ Locust Pose (Salabhasana) pages 94–95

10 Boat Pose (Paripurna Navasana) pages 110–111

11 Marichi's Pose (Marichyasana) pages 120–121

12 Bound Angle Pose (Baddha Konasana) page 104

13 Wide-Angle Seated Forward Bend (Upavisha Konasana) pages 72–73

14 Head-to-Knee Forward Bend (Janu Sirsasana) page 68

15 One-Legged KIng Pigeon Pose, easier (Eka Pada Rajakapotasana) pages 96–97

16 Bridge Pose (Setu Bandhasana) pages 86–87

17 Reclining Twist pages 116–117

18 Corpse Pose (Savasana) page 29

INTERMEDIATE SEQUENCE

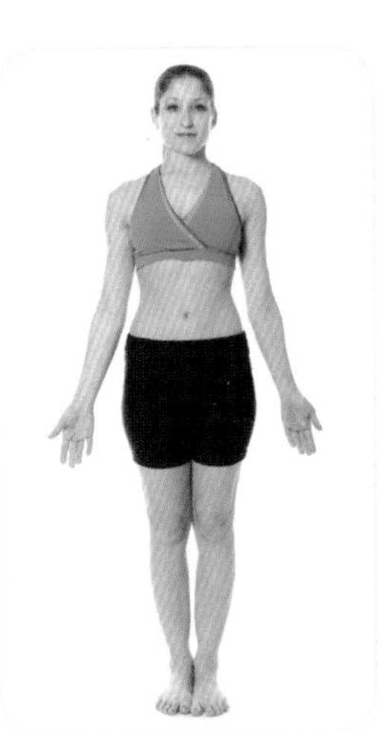

❶ Mountain Pose (Tadasana) page 32

❷ Twisting Chair Pose (Parivrtta Utkatasana) pages 124–125

❸ Garland Pose (Malasana) pages 34–35

❹ Crane Pose (Bakasana) pages 130–131

❺ Warrior II (Virabhadrasana II) pages 56–57

❻ Half Moon Pose (Ardha Chandrasana) pages 46–47

❼ Triangle Pose (Trikonasana) page 42

❽ Revolved Triangle Pose (Parivrtta Trikonasana) pages 44–45

❾ Plank Pose page 134

❿ Side Plank Pose (Vasisthasana) pages 138–139

⓫ Downward-Facing Dog (Adho Ukha Svanasana) page 24

⓬ Plank Pose page 134

⓭ Upward-Facing Bow Pose (Urdhva Dhanurasana) pages 88–89

⓮ Knees-to-Chest Pose (Apanasana) page 28

⓯ Supported Shoulderstand (Salamba Sarvangasana) pages 142–143

⓰ Plow Pose (Halasana) pages 140–141

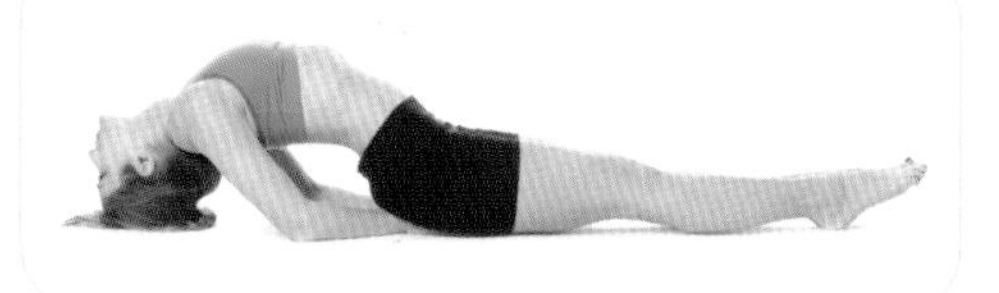

⓱ Fish Pose (Matsyasana) pages 92–93

⓲ Fire Log Pose (Agnistabhasana) page 105

⓳ Half Lord of the Fishes Pose (Ardha Matsyendrasana) pages 122–123

⓴ Corpse Pose (Savasana) page 29

ADVANCED SEQUENCE

❶ Easy Pose (Sukhasana) page 22

❷ Bharadvaja's Twist (Bharadvajasana I) pages 114-115

❸ Downward–Facing Dog (Adho Mukha Svanasana) page 24

❹ Plank Pose page 134

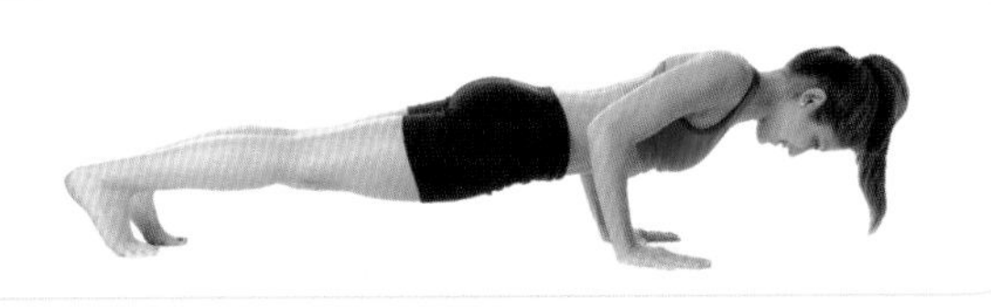

❺ Four-Limbed Staff Pose (Chaturanga Dandasana) page 135

❻ Downward-Facing Dog (Adho Mukha Svanasana) page 24

❼ Warrior II (Virabhadrasana II) pages 56–57

❽ Extended Side-Angle Pose (Utthita Parsvakonasana) pages 60–61

❾ Warrior I (Virabhadrasana I) pages 54–55

❿ Warrior III (Virabhadrasana III) pages 58–59

⑪ Hero Pose (Virasana) page 102

⑫ Reclining Hero Pose (Supta Virasana) page 103

⑬ Low Lunge Pose (Anjaneyasana) pages 50–51

⑭ Monkey Pose (Hanumanasana) pages 112–113

⑮ Half Lotus Pose (Ardha Padmasana) page 108

⑯ Full Lotus Pose (Padmasana) page 109

⑰ Eight-Angle Pose (Astavakrasana) pages 136–137

⑱ Supported Headstand (Salamba Sirsasana) page 144–145

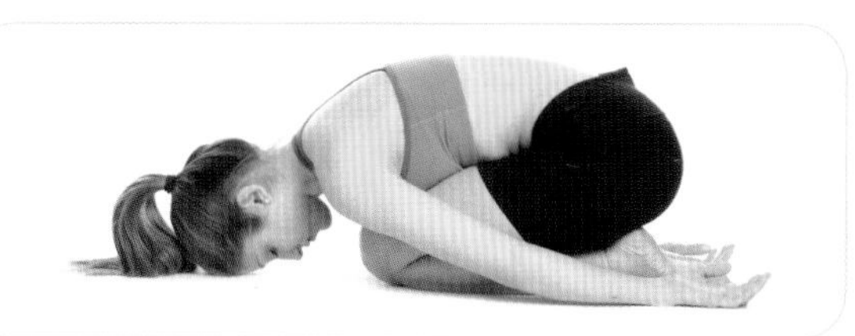

⑲ Child's Pose (Balasana) page 27

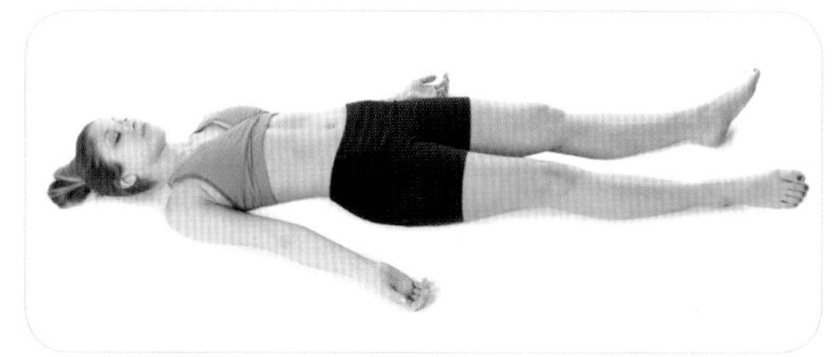

⑳ Corpse Pose (Savasana) page 29

MUSCLE GLOSSARY

The following glossary explains the Latin terminology used to describe the body's musculature. Certain words are derived from Greek; these have been indicated in each instance.

Neck

levator scapulae: *levare*, "to raise," and *scapulae*, "shoulder [blades]"
scalenes: Greek *skalénós*, "unequal"
splenius: Greek *splénion*, "plaster, patch"
sternocleidomastoideus: Greek *stérnon*, "chest," Greek *kleís*, "key," and Greek *mastoeidés*, "breast-like"

Back

erector spinae: *erectus*, "straight," and *spina*, "thorn"
latissimus dorsi: *latus*, "wide," and *dorsum*, "back"
multifidus spinae: *multus*, "much," *findere*, "to split," and *spina*, "thorn"
quadratus lumborum: *quadratus*, "square" or "rectangular," and *lumbus*, "loin"
rhomboideus: Greek *rhembesthai*, "to spin"
trapezius: Greek *trapezion*, "small table"

Chest

coracobrachialis: Greek *korakoeidés*, "raven-like," and *brachium*, "arm"
pectoralis [major and minor]: *pectus*, "breast"

Shoulders

deltoideus [anterior, posterior, and medialis]: Greek *deltoeidés*, "delta-shaped"
infraspinatus: *infra*, "under," and *spina*, "thorn"
subscapularis: *sub*, "below," and *scapulae*, "shoulder [blades]"
supraspinatus: *supra*, "above," and *spina*, "thorn"
teres [major and minor]: *teres*, "rounded"

Core

obliquus externus: *obliquus*, "slanting," and *externus*, "outward"
obliquus internus: *obliquus*, "slanting," and *internus*, "within"
rectus abdominis: *rego*, "straight, upright," and *abdomen*, "belly"
serratus anterior: *serra*, "saw," and *ante*, "before"
transversus abdominis: *transversus*, "athwart," and *abdomen*, "belly"

Hips

gemellus inferior: *geminus*, "twin" and *inferus*, "under"
gemellus superior: *geminus*, "twin" and *super*, "above"
gluteus maximus: Greek *gloutós*, "rump," with Latin *maximus*, "largest"
gluteus medius: Greek *gloutós*, "rump," with Latin *medialis*, "middle"
iliacus: *ilia*, "groin"

iliopsoas: *ilia*, "groin," and Greek *psoa,* "groin muscle"
iliotibial band: *ilia*, "groin," and *tibia*, "reed pipe"
obturator externus: *obturare*, "to block," and *externus*, "outward"
obturator internus: *obturare*, "to block," and *internus*, "within"
pectineus: *pectin*, "comb"
piriformis: *pirum*, "pear," and *forma*, "shape"
quadratus femoris: *quadratus*, "square" or "rectangular," and *femur*, "thigh"

Upper Arm

biceps brachii: *biceps*, "two-headed," and *brachium*, "arm"
brachialis: *brachium*, "arm"
triceps brachii: *triceps*, "three-headed," and *brachium*, "arm"

Lower Arm

brachioradialis: *brachium*, "arm," and *radius*, "spoke"
extensor carpi radialis: *extendere*, "to bend"; Greek *karpós*, "wrist"; and *radius*, "spoke"
extensor digitorum: *extendere*, "to bend," and *digitus*, "finger, toe"
flexor carpi radialis: *lectere*, "to bend"; Greek *karpós*, "wrist"; and *radius*, "spoke"
flexor digitorum: *flectere*, "to bend," and *digitus*, "finger" or "toe"

Upper Leg

adductor longus: *adducere*, "to contract," and *longus*, "long"
adductor magnus: *adducere*, "to contract," and *magnus*, "major"
biceps femoris: *biceps*, "two-headed," and *femur*, "thigh"
gracilis: *gracilis*, "slim, slender"
rectus femoris: *rego*, "straight" or "upright," and *femur,* "thigh"
sartorius: *sarcio*, "to patch" or "to repair"
semimembranosus: *semi*, "half," and *membrum*, "limb"
semitendinosus: *semi*, "half," and *tendo*, "tendon"
tensor fasciae latae: *tenere*, "to stretch"; *fasciae*, "band"; and *latae* "laid down"
vastus intermedius: *vastus*, "immense, huge," and *intermedius*, "that which is between"
vastus lateralis: *vastus*, "immense, huge," and *lateralis*, "of the side"
vastus medialis: *vastus*, "immense, huge," and *medialis*, "middle"

Lower Leg

extensor hallucis: *extendere*, "to bend," and *hallex*, "big toe"
flexor hallucis: *flectere*, "to bend," and *hallex*, "big toe"
gastrocnemius: Greek *gastroknémía*, "calf [of the leg]"
peroneus: *peronei,* "of the fibula"
soleus: *solea*, "sandal"
tibialis anterior: *tibia*, "reed pipe," and *ante*, "before"
tibialis posterior: *tibia*, "reed pipe," and *posterus*, "coming after"

CREDITS & ACKNOWLEDGMENTS

All photographs by Jonathan Conklin/Jonathan Conklin Photography, Inc.

Model: Zahava "Goldie" Karpel

All illustrations by Hector Aiza/3D Labz Animation India, except the insets on pages 16, 17, 18, 19, 35, 36, 39, 41, 43, 45, 47, 49, 51, 57, 59, 61, 65, 69, 71, 73, 79, 81, 83, 87, 89, 91, 93, 95, 102, 104, 105, 107, 108, 113, 119, 121, 123, 125, 129, 131, 133, 137, by Linda Bucklin/Shutterstock

Acknowledgments

The author and publisher offer thanks to those closely involved in the creation of this book: Moseley Road president Sean Moore; editor/designer Amy Pierce; art director Brian MacMullen; editorial director/designer Lisa Purcell; and assistant editor Jon Derengowski.